ASSESSING TOXIC EFFECTS
OF
ENVIRONMENTAL POLLUTANTS

ASSESSING TOXIC EFFECTS
OF
ENVIRONMENTAL POLLUTANTS

edited by

S. D. LEE

Research Toxicologist
Environmental Criteria and Assessment Office
U.S. Environmental Protection Agency
Cincinnati, Ohio

J. BRIAN MUDD

Professor of Biochemistry
University of California, Riverside

ANN ARBOR SCIENCE
PUBLISHERS INC
P.O. BOX 1425 • ANN ARBOR, MICH. 48106

Library of Congress Catalog Card No. 78-71430
ISBN 0-250-40266-1

Assessment of chemical toxicity has advanced far from the stages of determining the dose resulting in lethality. Even though this test of lethality provides necessary information, it is important to find other methods of assessment which are rapid and can provide the basis for predicting toxicity. Biophysical and biochemical methods are described in this book which may provide sensitive methods for assessing toxicity as well as discovering the underlying mechanisms by which toxic chemicals act.

The need for information on the toxicity of environmental pollutants is clearly based on the need to protect human health. The first four chapters of this book concentrate on human health, covering research with human subjects and nonhuman primates, and discussing the difficulties of assessing human toxicity on the basis of tests using laboratory animals.

Throughout the book, the emphasis is on the lung, because environmental pollutants are frequently inhaled. Included are discussions of the use of lung organ cultures for assessment of toxicity and the use of specific cell types of the lung, particularly those cells providing resistance to bacterial infection. It is important to know which regions of the lung are affected by inhaled pollutants. One chapter develops methods for predicting these regions based on the chemistry of the pollutant and degree of exercise.

Many pollutants from many sources are examined: primary and secondary pollutants from automobile exhaust, such as carbon monoxide and ozone; pollutants from industrial sources, such as sulfur dioxide; herbicides and pesticides; and cigarette smoke. The effects which are examined range from perturbations of the cellular membrane of the cell to the study of mutagenicity and carcinogenicity.

This book introduces material of interest to people of many disciplines of chemistry and biology who are concerned with deleterious chemicals in our environment. While reviewing progress in these areas of toxicology, new and specific research data are presented. This book should be considered as a companion volume to *Biochemical Effects of Environmental Pollutants*, S. D. Lee, editor, also published by Ann Arbor Science Publishers, Inc.

The chapters in this book are the outcome of a symposium at the 174th American Chemical Society Meetings in Chicago. The emphasis at the symposium, and in this book, was on the assessment of toxicity of environmental pollutants.

The editors are grateful to Dr. Linda Deans and Dr. Nina McClelland, Chairpersons, Division of Environmental Chemistry, American Chemical Society, for their support of the symposium which led to this book.

S. D. Lee
J. B. Mudd

 **Dr. S. D. Lee** is Research Toxicologist, Environmental Criteria and Assessment Office, U.S. Environmental Protection Agency, Cincinnati, and Adjunct Assistant Professor, Department of Environmental Health, University of Cincinnati College of Medicine.

He received his PhD from the University of Maryland, specializing in animal nutrition and biochemistry. He continued his training in biochemistry at Duke University Medical Center under a postdoctoral fellowship from the National Institute of Health. He was awarded an Advance Research Fellowship from the American Heart Association for continuation of his work at Duke.

His research work at EPA has been primarily devoted to the early identification of adverse effects of environmental pollutants using animal models, to obtain necessary information for assessing possible health effects on human populations. More recently, he has been engaged in health assessment and criteria document preparation of various water-associated pollutants.

Dr. Lee is the author of numerous papers presented at national and international conferences and symposia, and has published over forty articles in various professional journals. He is the editor of *Biochemical Effects of Environmental Pollutants*, published by Ann Arbor Science in 1977.

 **Dr. J. Brian Mudd** is Professor of Biochemistry at the University of California, Riverside. He received a Bachelor's Degree from Cambridge University, England, a Master's Degree from the University of Alberta, Canada, and a PhD from the University of Wisconsin. His research has concentrated on the biochemical effects of air pollutants, particularly ozone and peroxyacetylnitrate. This research has ranged from studies of these compounds with biochemicals in well-defined systems to the current studies on cells.

TABLE OF CONTENTS

CHAPTER 1

FROM ANIMALS TO MAN,
THE GRAND EXTRAPOLATION
OF ENVIRONMENTAL TOXICOLOGY

D. B. Menzel

Departments of Pharmacology and Medicine
Duke University Medical Center
Durham, North Carolina 27710

INTRODUCTION

There are few areas more important in the application of chemistry than environmental toxicology. To extrapolate from animal experiments to effects on man, a critical understanding of the mechanism of toxicant action is needed. Understanding the chemical mechanism of action allows the proper utilization of animal, plant and microorganism models for studying the potency and effect of a given environmental toxicant. Much of the toxicological data fails to meet this criterion. Much data is simply a repetition of highly standardized and regimented tests of individual compounds to provide an indication of its potency compared to other compounds. While there is no doubt that such screening tests are needed and must be continued, they add little to our knowledge of the overall problem of toxicity. Recent legislative action in the United States leading to the implementation of the Toxic Substances Hazard Act places us in critical shortage of personnel, resources and funds that will not allow us the luxury of examining case by case every compound that is likely to enter the market place or to be dispersed widely in the environment. It is my view that we must apply the techniques and technology of chemistry to an understanding of the toxicology of classes of compounds and do so on the basis of the molecular chemistry involved. Implicit in my argument will be the concept that abatement or prevention of exposure is the only safe strategy to control toxicants in the environment and work place. We know of no cure for cancer. We know of no cure for mutagenically related birth defects. Until these modes of therapeutic medicine become available, prevention remains the only cure.

It is the purpose of this chapter to stimulate thought on this problem through a fundamental approach in the hope that by applying this methodology we might predict toxicity. The following are several examples that illustrate both the problems facing the environmental toxicologist and the nature of the chemistry involved.

MUTAGENESIS, CARCINOGENESIS AND METABOLISM

It is proposed that much emphasis be placed on a hierarchical model for testing the very compounds falling under the Toxic Substances Hazard Act. Mutagenesis of such compounds is supposed to be detected by the reversion of specific traits of microorganisms. This test system, widely known as the Ames Test after its originator, Professor Bruce Ames, utilizes a simple bioassay system.[1] Mutants of the microorganism, *Salmonella typhimurium,* have been selected to lack specific traits for growth and survival in medium deficient of specific nutrients. Should the test organism be exposed to a chemical that causes a mutation, by chance some of the mutations will be reversions back to the genome which does not require the externally supplied nutrient for growth. It is then a simple task for the microbiologist to measure the incidence of such mutations by counting the number of revertent colonies after treatment with the chemical in question. The structure of the fire retardant Tris and several decomposition products are shown in Figure 1. Typical data are shown in Figure 2 taken from Dr. Ames' work,[2] in which the number of revertents per plate is directly proportional to the concentration of the potential mutagen that was applied to the test organism. Several decomposition products are not as active as the parent compound. These low mutation rates are typical of many compounds tested. The mutant microorganisms are selected so that they lack any reparative mechanism; thus, the mutation rate detected under these conditions represents the maximum possible. Ames and others have manipulated these and other kinds of microorganisms to reduce the concentration gradient that might exist between the medium and the interior of the microorganism. They have sought to have microorganisms that are freely permeable to a large number of complex organic compounds.

Often the chemical is not mutagenic in itself but requires metabolism to a transient but highly reactive intermediary, which appears to bind

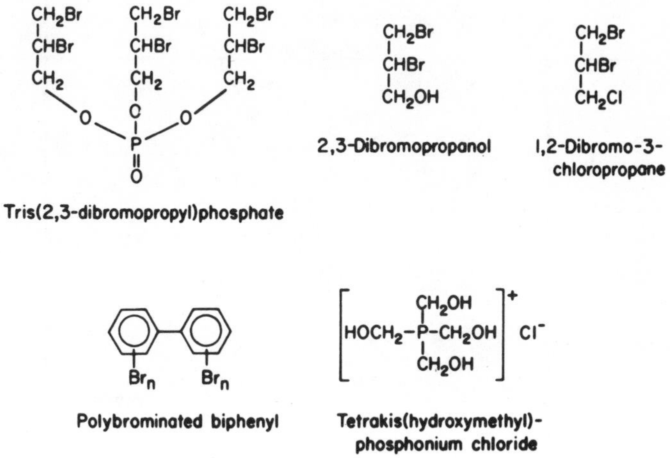

Figure 1. Chemical structure of the fire retardants and related compounds. (From Blum and Ames,[2] with permission of the authors and publishers.)

covalently to the nucleic acid or the macromolecules of the test organism. Activation in this system is accomplished by the addition of a crude microsomal preparation obtained from the liver of some mammal. Much of the activation appears to be carried out by the mixed function oxidases of the hepatic microsomal fraction called S9. NADPH and molecular oxygen are required. In the Ames system then, NADPH or an NADPH generating system is added to the microsomal fraction along with the test organism and the toxicant. The results in Figure 2 are in the presence of this fraction isolated from the liver of a rat induced by prior treatment with Arochlor.

A major unresolved question in the use of the Ames method of activation of foreign compounds is the chemical nature of the products formed. The chemical composition of most of the activated compounds found to be mutagenic has been inferred by the use of preformed metabolites thought to be the ultimate product of the microsomal oxidase. Arene oxides have been proposed as the most common intermediaries for polycyclic aromatic hydrocarbon-type compounds. Alkylation is assumed to be the most common mechanism of reaction between the intermediary toxicant and DNA. Until the chemical nature of the activated compounds reaching the microorganism is known, one cannot be sure of the ultimate utility of this test in terms of extrapolation to man. Purposefully, the Ames test has been rendered more sensitive by eliminating the soluble enzymes capable of detoxification of the "activated" metabolites formed

4

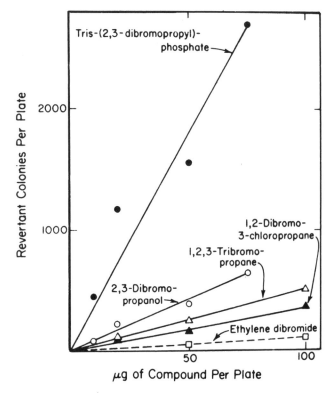

Tris-(2,3-dibromopropyl)-phosphate

2,3-Dibromo-propanol

1,2,3-Tribromo-propane

1,2-Dibromo-3-chloropropane

Ethylene dibromide

Revertant Colonies Per Plate

μg of Compound Per Plate

Figure 2. Number of revertant mutations produced in *Salmonella* strain TA100 by the fire retardant Tris and related compounds. The number of revertant colonies per plate is a measure of the number of mutations. Revertant colonies represent bacteria capable of growing in the histidine-deficient medium. Ethylene dibromide was added to the plates at 10 times the scale values. The results with tris-BP, 2,3-dibromopropanol and dibromochloropropane were obtained in the presence of the activating system from an Arocholor-treated rat. (From Blum and Ames,[2] with permission of the authors and publishers.)

by the microsomal oxidase. Little progress has been reported in this area, so one cannot be sure that a compound found to be mutagenic in the Ames system would be mutagenic in man. It may be rapidly detoxified and not have an opportunity to express its mutagenic potential. The kinetics of conversion and detoxification have not been fully worked out for model compounds, as is evident in the following discussion on the non-mutagenic toxic effects of covalent binding.

It is also clear that multiple forms exist[3] of the principal microsomal oxidase, cytochrome P_{450}. The expression of these isoenzymes depends on both the genome of the animal donating the microsomes and the environmental exposure of the animal. Sometimes, inducers of cytochrome P_{450}

are used prior to the preparation of the liver S9 fraction in the hope of increasing the activity of the preparation and, therefore, the sensitivity of the test. Since the products of the monooxygenase remain elusive, one cannot be sure of the chemical effect resulting from the use of an S9 fraction containing different isomeric forms of cytochrome P_{450}. Again, much progress can be gained from the chemical identification of the products formed using these different isoenzymic cytochrome P_{450} preparations.

The chemical nature of the reaction between the target within the microorganism and the "activated" mutagen is also not clearly understood; nor, for that matter, is the target macromolecule fully known. In more complex eukaryotic cells, reaction with macromolecules other than nucleic acids may result in mutagenesis through complex mechanisms such as reverse transcriptase or alterations in local membrane properties. False negative tests could result from such imperfections in the microorganism-based test.

A major problem exists in the question of a "threshold" or "no effect" dose for mutagens. Certainly, the concentration and frequency of naturally occurring mutagens and carcinogens in plants and foods are greater than those thought to occur from the apparent mutation and cancer rates in man. Since we do not know the pharmacokinetics of the absorption, distribution, activation and elimination of these mutagens, one cannot presume an inherent ability of man to repair mutagenesis at certain low doses of mutagens or carcinogens. The existence of a "no effect" dose for a given carcinogen may be due to an artifact of the kinetics of these reactions unique to that carcinogen and not to a general phenomenon.

Lastly, the heart of the hypothesis that carcinogenesis is due to mutagenesis remains to be proved. One must admit that the prediction thus far by microorganism-based tests is impressive and that all carcinogens have proved to be mutagens. Not all mutagens have proved to be carcinogens, however. Quite properly, this fact has led to the development of a hierarchical approach, wherein the microbiological test is the first tier. Unfortunately, there is a growing tendency to stop at the Ames test once the compound has been found mutagenic. Likewise, the Ames test does not predict any other toxic reactions. The pulmonary toxicity of the herbicide paraquat could not be predicted from the Ames test and represents to the occupationally exposed as great a hazard as a potential malignancy.

Metabolic activation to reactive intermediaries, which combine covalently with cell constituents, is a general toxic reaction not confined to mutagens or carcinogens. The pioneering work[4] in demonstrating that drugs and other toxicants may react covalently through activation by a microsomal system with tissue constituents *in vivo* is illustrated with acetaminophen.

Acetaminophen induces hepatic necrosis when given above a critical concentration. The time course of covalent binding of tritiated acetaminophen to mouse tissues has been measured by Jollow, et al.[3] Covalent binding of acetaminophen to the liver occurs rapidly while muscle tissue has very little, if any, binding. Acetaminophen is then rapidly metabolized in the liver to an intermediary covalently bound to tissue macromolecules, which is only slowly removed during 24 hours after the administration. The maximum covalent binding of acetaminophen under these studies, which is typical of the metabolism of drugs that bind covalently to the liver, is only about 2 nmol/mg of tissue protein. In the conventional balance table approach to studies of distribution and uptake of drugs and toxicants, such binding would represent only a miniscule amount, less than 0.01% of the total dose given. The detection of such amounts is unlikely without high specific activity radiolabeled compounds.

Activation of compounds leading to tissue alkylation is presumed to proceed via arene oxides. Other pathways that prevent tissue alkylation by detoxification, such as hydration and rearrangement, exist in mammalian tissues. Conjugates with glutathione may or may not represent full detoxification. Glutathione is a good leaving group, and subsequent displacement reactions on the conjugate can occur.

A critical problem is to model this more complex balance of activating and detoxifying reactions in a system such as the Ames test. At present, we can only presume that any mutagen detected by the microbial system is a potential carcinogen.

Mutation need not lead to malignant transformation. Mutation could lead to a selective advantage on the part of the daughter cells, promoting a benign growth. The monoclonal nature of atherosclerotic plaques has led to the suggestion that this chronic disease might be due to a single mutation in the fibrocytes lining the arteries.[5] Other chronic diseases having their expression years after the initiating event could also be the result of a single mutation, which is propagated slowly by the development of clones within an organ system. Degenerative diseases of a monoclonal nature may then be caused in part by environmental mutagens.

HORMONAL PATHWAYS OF TOXICITY

Air pollutants react with tissue constituents and thereby may produce "ultimate toxicants" in a manner analogous to mutagenic and carcinogenic materials. Ozone and nitrogen dioxide react readily with a variety of model molecular species to produce reactive compounds.

Unsaturated fatty acids are particularly susceptible to attack yielding peroxides, ozonides and aldehydes, which have systemic effects in organs other than the lung.

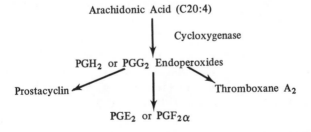

Initiation of Peroxidation
of Tissue Unsaturated Lipids

Initiation of Peroxidation of Tissue
Unsaturated Lipids

Dietary effects also occur. Vitamin E deficiency increases the toxicity of ozone and nitrogen dioxide.[6] For instance, sleeping time of mice due to pentobarbital injection is elongated by prior exposure to ozone, suggesting a direct effect on the liver.[7]

Hormonal-like activity is produced by ozone-catalyzed autoxidation of arachidonic acid similar to the enzyme-catalyzed oxygenation, which is the initial step in the formation of prostaglandin and related hormones.

Arachidonic Acid (C20:4)

Cycloxygenase

PGH_2 or PGG_2 Endoperoxides

Prostacyclin   Thromboxane A_2

PGE_2 or $PGF_{2\alpha}$

Hormonal activity, similar to prostaglandin endoperoxides PGG_2 and PGH_2, has been found using assays with human platelets.[8] Human platelets are rapidly aggregated by arachidonic acid endoperoxides formed during the ozone autoxidation of arachidonic acid. Aggregation is indicated in Figure 3 by the downward deflection due to a change in

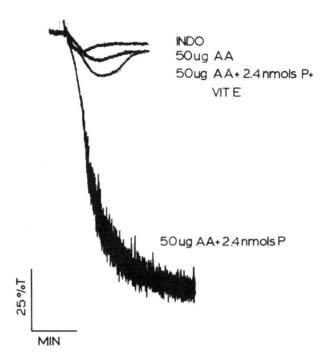

INDO
50ug AA
50ug AA + 2.4 nmols P+
VIT E

50ug AA + 2.4 nmols P

25%T

MIN

Figure 3. Aggregation of human platelets by arachidonic acid peroxides illustrating the prostaglandin-hormone-like activity. Arachidonic acid (AA) or AA with vitamin E (1:30^3 molar ratio of AA to vitamin E) was peroxidized by a stream of 2.0 ppm ozone for 1 minute. Platelets were aggregated by AA, arachidonic acid peroxides (AAP) formed with the presence or absence of vitamin E. Platelets were or were not preincubated with 14 μM indomethacin (Indo), an inhibitor of platelet cycloxygenase. (From Roycroft, *et al.*,[8] with permission of the authors and publishers.)

turbidity of a suspension in human platelet-rich plasma. Vitamin E, a required nutrient for man and animals, has a profound effect on the biological potency of the peroxides formed. Vitamin E abolishes most of the biological activity while not inhibiting the formation of peroxides. The ozone-formed peroxides also contracted aortic spiral strips and fundus strips, demonstrating biological activity on smooth muscles nearly as potent as the natural endoperoxides, PGG$_2$ and PGH$_2$.

A scheme for the formation of cyclic peroxides during the autoxidation of cellular unsaturated fatty acids is shown in Figure 4. Toxicity through lipid peroxidation begins with the abstraction of a hydrogen from the *cis*-methylene-interrupted fatty acids found in all cells. Fatty acids having

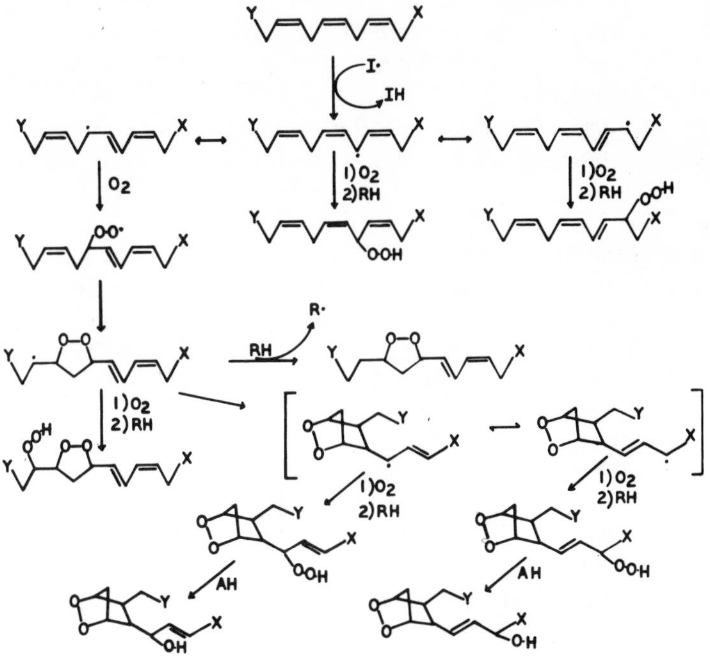

Figure 4. A scheme for the formation of cyclic peroxides having hormone-like activity on the peroxidation of unsaturated fatty acids. X and Y represent the carboxylic and alkyl residues of the fatty acid. Peroxidation is initiated by a free radical I·Hydrogen abstraction from another fatty acid molecule RH provides another radical R· for propagation of the reaction. (From D. B. Menzel, "Environmental Toxicants and Vitamin E" in *Vitamin E*, L. Machlin, Ed., (New York: Marcel Dekker Press, (In Press), with permission of the publishers.)

three or more unsaturations are of particular interest. One resonance form can occur that will result in the formation of a peroxyl free radical β-γ to an unsaturation. Such a peroxyl group can cyclize to form an endoperoxide. This reaction path is thought to be that of the enzyme cycloxygenase, which forms PGH_2 and PGG_2 from arachidonic acid. Prostanoic acid can result from such a cyclization and has been isolated from peroxidizing polyunsaturated fatty acids. Nonenzymatic and enzymatic conversion of unsaturated fatty acids having greater than three unsaturations, and hence the possibility of a β-γ unsaturated peroxyl free radical, results in the same prostaglandin product. Under this scheme, the end effect is independent of the nature of the initiating compound. Peroxidation could come from inhaled ozone (O_3) and nitrogen dioxide (NO_2) or from the metabolic activation of ethanol or carbon tetrachloride. All of the exposures result in free radicals capable of initiation. The organ most affected will be that receiving the greatest dose of the toxicant. The organ response will differ depending on the nature of the potential "prostaglandin-like" products formed.

The final endoperoxides formed by ozone autoxidation, no doubt, are complex mixtures of different stereoisomers and only a very minor fraction is likely to be identical with the enzymatically formed PGH_2. Nonetheless, they have potent biological activity. The central role of prostaglandins in the homeostasis of the circulation and blood vessels may mean that ozone-catalyzed reactions such as these or lipid peroxidation catalyzed by drug or toxicant intake may lead to atherosclerosis, stroke and cardiovascular disease. Lipid peroxidation is known to be initiated by a wide variety of compounds aside from air pollutants such as carbon tetrachloride, ethanol and several halogenated hydrocarbons.

Dietary factors such as vitamin E have profound effects on both the end products formed from ozone-catalyzed autoxidation and the morbidity of continuous O_3 and NO_2 exposure. Vitamin E reacts readily with peroxyl free radicals but only slowly with alkyl free radicals. Alkyl free radicals, on the other hand, react rapidly with molecular oxygen. Vitamin E promotes the formation of hydroperoxides and decreases the formation of cyclic peroxides. Hydroperoxides are much less active in the prostaglandin system and are easily degraded by glutathione peroxidase. The protection afforded by vitamin E ingestion might be explained by the overall termination of the free radical chain reaction leading to lipid peroxidation. The specificity of vitamin E in biological systems may lie with its association with membrane lipids in such a way as to promote hydroperoxide formation over cyclic peroxides. Hydroperoxides, being less active as hormones and better as substrates for glutathione peroxidase, are part of a potential detoxification pathway.

In man, dietary patterns vary considerably and have specific ethnic trends. In the United States, for example, the intake of vitamin E is low. Freezing does not preserve vitamin E well, due to the free radical nature of its decomposition reactions. The consumption of frozen foods promotes a low vitamin E intake. There is also a trend toward more unsaturated fatty acids in the diet for cardiovascular protection. These effects may exacerbate the toxicity of peroxidizing toxicants. Fatty acid hydroperoxides are also potent inhibitors of the enzyme synthesizing prostacyclin. Prostacyclin is a potent antagonist of the prostaglandins and thromboxanes. Thromboxanes promote platelet adhesion, thought to be the initial step in arterial injury leading to embolism. Ironically, the strategy proposed to decrease cardiovascular disease may promote increased toxicity for air pollutants.

While these observations on the toxicity of air pollutants as modified by diet have direct implications for man, they also illustrate a facet hardly ever taken into account in comparative toxicology. Diets can alter by several orders of magnitude the response of animals to a toxicant. Little control is exercised over animal diets, nor are they made to mimic human dietary regimens. Few animals are rendered deficient or marginal in vitamins or minerals known to be limiting in the free living population. Aside from low vitamin E intake, the United States population also has a marginal intake of bioavailable iron, riboflavin and folic acid. Specific groups within the general population may also have low calcium and trace mineral intakes as well. Much knowledge about these marginal intakes is available by geographic area, age, sex and race. These deficiencies are not so frank as to be classical vitamin or mineral deficiencies; yet from a biochemical viewpoint, the marginal intake limits severely the ability of the patient to cope with toxicants. A definite need exists to model these marginal nutritional states for particular segments of the population, such as the neonate, the growing child, the pregnant teenager, the pregnant adult and the elderly.

Toxicity is never tested on animals having preexisting disease. But a sizable fraction of our population already has preexisting disease, possibly as a result of a chronic and unsuspected exposure to some environmental toxicant. Models of human disease in animals are limited, but even those that have been established have not been used for toxicity studies. Alloxan diabetes in the rat simulates human diabetes, for example, and could be used to study the potential interactions of this common genetic disease of man with specific toxicants.

Toxicants may interact in other ways with hormonal receptors to provide a chemical amplification of the toxic effect. A very well-known example of such interaction is the inhibition of acetylcholinesterase by organophosphate and carbamate insecticides. In the absence of cholinesterase activity, acetylcholine accumulates in neuronal synapses to produce a cholinomimetic action. Because of the high turnover number of cholinesterase, inhibition of a single cholinesterase enzyme molecule results in an amplification of that event several thousand times. Other ways in which hormonal interactions can result from toxicant exposure are indicated by recent results in our laboratory on the binding of concanavallin A, wheat germ agglutinin, wheat germ lipase and immunoglobins to plasma membrane-bound receptors of pulmonary macrophages isolated from rabbits exposed *in vivo* to either nitrogen dioxide or ozone.[9]

Exposures as short as 3 hr to 0.5 ppm ozone or 24 hr to 7.0 ppm NO_2 augment the rosette formation between red cells and macrophages mediated by wheat germ agglutinin. A detailed analysis of the binding of these ligands to their respective receptors on the surface of the membrane does not demonstrate any appreciable change in the intrinsic affinity of the receptor for the ligand. These alterations may represent changes in the fluidity of the membrane or in the topographical distribution of the receptors changing some of the "covert" receptors to "overt" ones. Aggregation of surface receptors due to disturbances in the fluidity of the membranes is difficult to detect other than by complicated assays of those reported here. However, hormonal ligands result in profound chemical amplification via receptor transducers such as adenylcyclase, so that small numbers of toxicant reaction sites can result in large biological actions resembling the native pathways. Membrane fluidity pathology is now recognized in man in derangements of the red blood cells. In Chapter 15 the activation of the histamine hormonal system by sulfate aerosols is reported as another example.

Air pollutants may also alter the ability of the host to protect itself against pathogens (Chapter 6). This mechanism of toxic action may be important in chronic exposures of man. To this end, studies of mouse model systems of infectivity are highly important. In this area, the basic tenet of toxicology of a dose-response relationship is in question. The effects of nitrogen dioxide exposure on the infectivity of murine viruses and bacteria may be more related to the absolute magnitude of the nitrogen dioxide concentration than to the product of nitrogen dioxide concentration times the time of exposure. These observations bring into question the conventionally accepted, but empirical, relationship of the dose-response curve. A true threshold or "no effect" level may not exist. Since the exposure of man to nitrogen dioxide is not at a constant level, but cyclical in nature, the results of these experiments are highly important.

SYNTHESIS OF PREDICTIVE MODELS

Lastly, measurement of the tissue dose of many toxicants is not technically feasible at present. An alternative approach is the development of mathematical models based on chemical reactions, rate equations, mass transport and anatomical information. Dr. Miller presents a specialized approach in terms of the inhalation toxicology of ozone in Chapter 14. Using morphological information of the mammalian lung, a breath-by-breath model of the distribution of reactive gases such as ozone has been modeled in animal and human airways. The absorption of ozone by lung tissue has been modeled by interposing an intermediary chemical reaction zone and by estimation of the removal of ozone by the nasopharyngeal cavity. Using such a model, a correlation was found between the

morphological lesion produced by ozone at the terminal respiratory bronchiole and the predicted uptake of ozone at this site in the lung. In other
words, the model predicted injury at the anatomical site most frequently
observed as damaged in experimental animals exposed to ozone. The
model can be manipulated to provide corrections of the dose of ozone
absorbed by different animal species at a given ambient concentration of
ozone and to compare the dose predicted to that absorbed in man. The
uptake of ozone may differ by two to three times between rabbits, guinea
pigs and man. The effects of lung pathology in man on the rate of absorption of ozone can likewise be modeled using this approach. Estimates
of health effects from animal exposure are thus more readily applicable
to actual human exposures and to specialized groups within the population
that may be at risk due to preexisting disease. The application of such
mathematical modeling is a general phenomenon that can well be applied
to other toxicants aside from ozone.

DISCUSSION AND CONCLUSIONS

Environmental toxicology must develop a hierarchy of models incorporating in each as much detailed chemical knowledge of the mechanism
or mechanisms of action of each toxicant as is known. The chemical
reactions are as diverse as the toxicants themselves. Simple models can
lead to more complex ones that can be generalized to classes of toxicants.
Because of the lifetime exposure to some toxicants and the prolonged
time required for effects to be observed in man, prediction today of future
intoxication is essential. Traditional concepts of anatomical pathology as
indices of intoxication must give way to a more molecular concept of
cell biology. A catalog of toxicants is not adequate since it can never be
predictive. A critical unresolved question is the dose-response relationship
that involves toxicant concentration, exposure time and the expression
of the toxic response. Even simple mathematical models of toxicity are
not available. Here the concepts of threshold and adaptation need reexam-
ination. The multiplicity of toxicants does not allow a case-by-case evaluation; rather, generalization is needed leading to prediction. Prevention of
exposure at present remains the most viable strategy for public protection.

ACKNOWLEDGMENTS

This work was supported by NIH Grants RO1 HL16264 and RO1
ES00798 and by EPA Contract 68-02-2436.

REFERENCES

1. Ames, B. N., S. D. Lee and W. E. Darston. "An Improved Bacterial Test System for the Detection and Classification of Mutagens and Carcinogens," *Proc. Nat. Acad. Sci., US* 70: 782-786 (1973).
2. Blum, A., and B. N. Ames. Flame-Retardant Additives as Possible Cancer Hazards. Science 195: 17-23 (1977).
3. Jollow, D. J., J. R. Mitchell, W. Z. Potter, D. C. Davis, J. R. Gillette and B. B. Brodie. "Acetaminophen-Induced Hepatic Necrosis. II. Role of Covalent Binding *in Vivo*," *J. Pharm. Exp. Therapeut.* 187: 195-202 (1973).
4. Haagen, D. A., M. J. Coon and D. W. Nebert. "Induction of Multiple Forms of Mouse Liver Cytochrome P450," *J. Biol. Chem.* 251: 1817-1826 (1976).
5. Ross, R., and J. A. Glomset. The Pathogenesis of Atherosclerosis. I and II," *New England J. Med.* 295: 369-377; 420-425 (1976).
6. Menzel, D. B., J. N. Roehm and S. D. Lee. "Vitamin E: The Environmental and Biological Antioxidant," *J. Food Agric. Chem.* 20: 481-486 (1972).
7. Gardner, D. E., J. W. Illing, F. J. Miller and D. L. Coffin. "The Effect of Ozone on Pentobarbitol Sleeping Time in Mice," *Res. Comm. Chem. Pathol. Pharm.* 9: 689-700 (1974).
8. Roycroft, J. H., W. B. Gunter and D. B. Menzel. "Ozone Toxicity: Hormone-Like Oxidation Products from Arachidonic Acid by Ozone-Catalyzed Autoxidation. *Toxicol. Lett.* 1: 75-82 (1977).
9. Hadley, J. G., D. E. Gardner, D. L. Coffin and D. B. Menzel. "Enhanced Binding of Autologous Red Cells to the Macrophage Plasma Membrane as a Sensitive Indicator of Pollutant Damage," in *Pulmonary Macrophage and Epithelial Cells*, D. L. Sanders, R. P. Schneider, G. E. Dagle and H. A. Ragan, Eds. (Springfield, VA: Technical Information Center, 1977), pp. 66-77.

CHAPTER 2

THE ROLE OF NONHUMAN PRIMATES IN ENVIRONMENTAL POLLUTION RESEARCH

W. Castleman, J. Gillespie, P. Kosch, L. Schwartz and
W. Tyler

> California Primate Research Center
> Departments of Veterinary Pathology and
>   Physiological Sciences
> University of California
> Davis, California 95616

INTRODUCTION

In discussing the role of nonhuman primates in environmental pollution research, this chapter will address, more specifically, the methods in which nonhuman primates have been used to assess potential public health hazards resulting from exposure to ambient levels of ozone and high levels of sulfuric acid aerosols.

Previous use of primates as experimental models has provided unique insight into a number of disease conditions affecting man. Examples are numerous of human diseases in which research with primates has produced invaluable understanding of the pathogenesis of those disease processes and vary from infectious diseases, such as viral hepatitis and kuru,[1,2] to drug-induced congenital malformations, such as those resulting from thalidomide administration.[3,4] It is our contention that nonhuman primates can also play an important role in assessing potential hazards to man resulting from environmental air pollutants.

In recent years, research with primates has suffered from the constraints of an ever-decreasing availability of animals. Numbers of primates imported for research have been severely restricted due to overall decreases in the world supply and limits imposed by exporting countries on numbers of wild primates which can be legally trapped and shipped to other countries. Greater emphasis is being placed by the National Institutes of Health on increasing domestic production of primates in breeding colonies. Associated with increased use of colony-bred primates has been an improved experimental animal at a greatly increased cost. Experimental studies with nonhuman primates in the past have often presumed that they are ideal models of man when there has often been few detailed comparative data to support that presumption.[5] Decreased supply and increased cost of primates now require a closer scrutiny of normal structural and functional aspects of organ systems in prospective primate models and comparison with available information for man. In other words, it is necessary to demonstrate whether primate models offer biologically important advantages over greatly more abundant and less expensive small laboratory animals, such as rats, mice and rabbits, or abundant, but still expensive animals, such as dogs and cats.

USE OF PRIMATES IN STUDIES WITH
OZONE AND SULFURIC ACID AEROSOL

Although there are difficulties in using primates, their use in research for evaluating the pulmonary effects of ozone has introduced what we feel are valuable concepts relative to potential dangers posed for humans living in ozone-polluted areas. Before discussing studies with primates, however, it would be useful to review briefly pulmonary lesions induced in rats by short-term exposure to near ambient levels of ozone. The term "near ambient" refers to levels of ozone below 1.0 parts per million (ppm). A number of studies have demonstrated that rats develop lesions in terminal bronchioles and in alveoli of proximal alveolar ducts following exposure to ozone levels of 0.9 ppm and below.[6-10] Following exposure to 0.8 and 0.5 ppm ozone for seven days, lumina of alveolar ducts and outpocketing alveoli are covered by clusters of inflammatory cells which are predominantly composed of macrophages. There is also hyperplasia of type II alveolar epithelial cells which most probably follows early damage of type I alveolar epithelial cells by ozone.[8,9] Terminal bronchiolar lesions are characterized by early damage to ciliated cells and proliferation by nonciliated bronchiolar epithelial cells.[8,11,12] There are also multifocal alterations in ciliated cells in more proximal intrapulmonary airways.[10] An important finding by Schwartz et al.[10] was that mild inflammatory lesions result in proximal areas of alveolar ducts following exposure for seven days to levels of ozone as low as 0.2 ppm. The implications of this finding are considerable because humans living in heavily polluted urban areas are frequently exposed to this or higher levels of ozone.[13]

An obvious question to raise is whether there is any similarity between lesions induced in rats by low levels of ozone and those that potentially could occur in humans. Also, is man as sensitive to ozone as are rats? Results from pulmonary function tests on humans exposed to low levels of ozone demonstrate functional changes after short periods.[14-17] The precise mechanisms underlying these functional changes and whether they are associated with structural lesions are not known.

This is the point at which nonhuman primates potentially can provide the most useful information. If there were major differences in the response of nonhuman primates and rats to ozone, it could as well be expected that there might be similar differences in the response of man. Similarities in responses, although not proving that lesions result in humans, would at least lend greater confidence to the significance of studies in rats. Hypotheses resulting from studies in rats could also be modified to encompass observations which were slightly contrasting in studies with primates.

Before ozone exposures in monkeys were begun, however, it was necessary to obtain additional information on normal pulmonary structure and function in primates. This was necessary initially because little information was available on pulmonary ultrastructure or pulmonary mechanics in primates. Even more importantly, since a presumption in starting these studies was that the response to ozone in nonhuman primates would be closely comparable to that of man, it was necessary to determine the degree of similarity between normal pulmonary structure and function in nonhuman primates and in man. In other words, are there any actual reasons based on pulmonary structure and function to suspect that monkeys might be a more suitable experimental model for man than rodents?

Pulmonary Airway Morphology in Macaques

Since studies in rodents have shown that distal pulmonary airways and immediately associated alveolar parenchyma in the proximal portion of the pulmonary acinus are the sites of the most severe ozone-induced lesions, our studies on normal pulmonary ultrastructure in primates concentrated most closely on distal airways and their epithelial lining. The species of most interest were macaques because of their relatively greater availability and the abundant biological data collected on them as a result of their past heavy use in biomedical research.

Rhesus, stumptail and bonnet monkeys (*Macaca mulatta, Macaca arctoides* and *Macaca radiata*, respectively) were the species studied initially by correlated methods of light microscopy, scanning electron microscopy and transmission electron microscopy.[18] It was found that macaques have a well-developed bronchial tree similar to that in humans[19] with a pseudo-stratified columnar epithelium composed primarily of ciliated cells, mucous cells and basal cells.[20] Nonrespiratory bronchioles are greatly abbreviated in macaques as compared to those in man. Macaques generally have a short, single-generation, terminal bronchiole distal to the end bronchus. In contrast, man has three or four generations of nonrespiratory bronchioles interposed between end bronchi and respiratory bronchioles. The bronchiolar epithelial lining in macaques is very similar to that in bronchi with mucous cells, ciliated cells and basal cells being predominant. In distal regions of terminal bronchioles in bonnet and stumptail monkeys, nonciliated bronchiolar cells (Clara cells) are present. Mucous cells are reduced greatly in number. Rhesus monkeys differ in that they do not have appreciable numbers of nonciliated bronchiolar epithelial cells in terminal bronchioles. Respiratory bronchioles are well developed in macaques and usually form at least two generations proximal to initial branches of alveolar ducts. Both the general structure and number of generations of respiratory bronchioles are similar between man and macaques.[19,21] In macaques, the ciliated epithelial surface ends in distal areas of terminal bronchioles. Respiratory bronchioles are lined by nonciliated epithelium composed of cuboidal cells arranged in small clusters or singly between squamous cells (Figure 1). The squamous cells appear to be identical to type I alveolar epithelial cells. The nonciliated cuboidal cells usually do not contain secretory droplets and appear to be a distinctive intermediate cell type between nonciliated bronchiolar epithelial cells in terminal bronchioles and type II epithelial cells (Figure 2). The mixture of squamous and cuboidal cells found in respiratory bronchioles of macaques has also been described in human respiratory bronchioles.[19,21] An important point to emphasize is that whereas both monkeys and man have well-developed respiratory bronchioles, rats and most other small laboratory animals do not. In these latter animals the pulmonary acinus is more simplified, and terminal bronchioles open into alveolar ducts.

Two other species of animals that have well-developed respiratory bronchioles are the dog and cat. The epithelium lining their airways differs from respiratory bronchiolar epithelium in the macaque in that there is a continuous lining by cuboidal cells (Figure 3) with no interspersing squamous cells. Also, the cuboidal cells contain abundant cytoplasmic deposits of glycogen, especially in the dog (Figure 4).

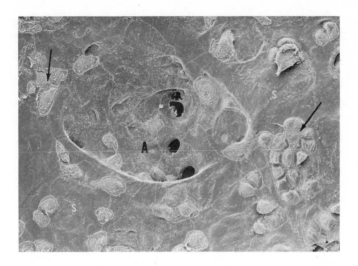

Figure 1. Scanning electron micrograph of respiratory bronchiole from a normal bonnet monkey. Cuboidal respiratory bronchiolar epithelial cells are present in clusters (arrows) between type I epithelial cells (s). A shallow alveolus is present in the center of the field (A) and has several interalveolar pores (X516).

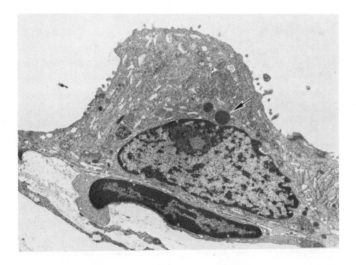

Figure 2. Transmission electron micrograph of a cubiodal respiratory bronchiolar epithelial cell from a normal bonnet monkey. Cytoplasm contains moderate numbers of mitochondria, abundant rough endoplasmic reticulum, and several perinuclear lysosome-like inclusions (arrow) (X9,450).

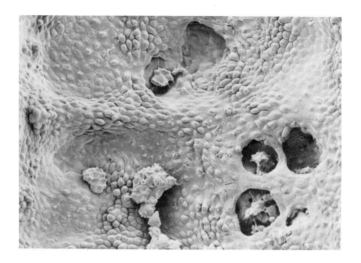

Figure 3. Scanning electron micrograph of respiratory bronchiole from a cat. There are several alveolar outpocketings. The wall is lined completely by cuboidal cells. Several clusters of macrophages are on the surface (arrows) (X540).

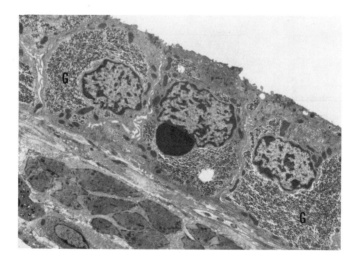

Figure 4. Transmission electron micrograph of cubiodal respiratory bronchiolar cells from a normal dog. The cytoplasm of these cells contains abundant glycogen (G) . The central perinuclear electron-dense material is probably lipid.

Pulmonary Function in Macaques

Another reason for choosing macaques for studies with ozone is that their size is more suitable than that of most other readily available species of nonhuman primates for detailed studies of pulmonary function. Physiological measurements were made describing the static and dynamic mechanical properties of the respiratory systems of rhesus and bonnet monkeys[22,23] and were compared with similar measurements available for man and common laboratory animals such as dogs and rodents. Anesthetized monkeys were studied while seated suspended upright in a whole-body plethysmograph. Subdivisions of lung volume and quasistatic volume-pressure curves of lung and chest wall were measured on 22 normal bonnet and 12 rhesus monkeys. Expiratory flow-volume curves and total pulmonary resistance were measured in 15 normal bonnet and 8 rhesus monkeys.

The most important findings from static pulmonary measurements were that monkey chest walls are about four times as stiff and their lungs nearly twice as compliant as those of beagle dogs.[24] Mean quasistatic deflation volume-pressure curves of lung and chest wall in 34 bonnet and rhesus monkeys and in 22 beagle dogs are shown in Figure 5. The stiff

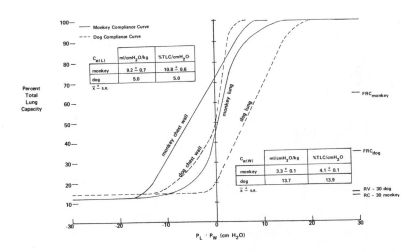

Figure 5. Mean quasistatic deflation volume-pressure diagrams of lung and chest wall in 34 bonnet and rhesus monkeys (——) and in 22 beagle dogs (– –): P_L = Transpulmonary pressure; P_W = transthoracic pressure; FRC = functional residual capacity; $C_{ST}(W)$ = chest wall compliance; $C_{ST}(L)$ = lung compliance; TLC = total lung capacity.[22]

chest wall of the monkey sets its functional residual capacity at a much larger percentage of total lung capacity (64%) than that of the beagle. dog (35%).[24] Values for man range from 50-54%.[25,26] Chest wall compliance relative to body weight is much greater in small rodents. For example, the mouse has an infinitely high chest wall compliance at functional residual capacity, which is only about 17% of its total lung capacity.[25] Measurements of expiratory flow-volume curves in these macaques show that the configuration of their maximum expiratory flow-volume curves is much more comparable to that of normal young adult humans than that of dogs (Figure 6).

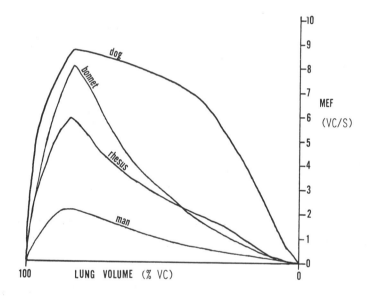

Figure 6. Maximum expiratory flow-volume curves for dogs, bonnet and rhesus monkeys, and man: MEF = maximum expiratory flow (expressed in vital capacities per second). Lung volume is expressed as percent vital capacity.[23]

The similarity of distal airway morphology as well as respiratory mechanical properties of macaques to those of man indicates the suitability of the macaque as an animal model in studies designed to evaluate potential effects of air pollutants on human pulmonary structure and function.

Ultrastructural Lesions Induced in Macaques by Ambient Levels of Ozone

In two groups of experiments, macaques were exposed to ambient and near ambient levels of ozone for seven days and pulmonary tissue studied by correlated light microscopy and scanning and transmission electron microscopy.[27-29] Twenty-four rhesus monkeys were used in the first study,[28] which examined effects of ozone at 0.8 and 0.5 ppm. This first study demonstrated that consistent lesions were produced at these higher ambient levels. Since one important objective of our studies was to determine whether primates are comparable to rats in their sensitivity to ozone, the second study using 11 bonnet monkeys[27,29] focused on effects of ozone at 0.35 and 0.20 ppm for seven days. Again, 0.2 ppm is the lowest level of ozone that has been shown to produce consistent lesions in proximal acinar areas of rats.[10] Changes in pulmonary structure of macaques were similar in nature and distribution, although not in severity, for the four levels of exposure. Because the findings in bonnet monkeys at 0.35 and 0.2 ppm ozone are likely to be more pertinent to the situation faced by man, they will be discussed in detail.

Four bonnet monkeys were exposed for seven days to 0.2 ppm ozone and three were exposed to 0.35 ppm as previously described.[29] Four monkeys exposed to filtered ambient air served as controls. Pulmonary lesions were present in all animals exposed to ozone. The damage was localized to pulmonary airways and was present in two anatomically related gradients of severity. One gradient was present in upper conducting airways and the other was present in respiratory bronchioles. In tracheas and proximal generations of bronchi most frequently and, to a lesser extent, in more distal conducting airways, there were clusters and longitudinal tracts of ciliated cells with greatly shortened and less numerous ciliary shafts (Figure 7). Also present in the same areas were increased numbers of nonciliated cells with apically located collections of centrioles. These latter cells were interpreted as being regenerating ciliated cells. The changes in the ciliated epithelial surface were much less severe and more variable in animals exposed to 0.2 ppm than those to 0.35 ppm. The most severe and consistently produced pulmonary lesions following exposure to either level of ozone, however, were in proximal areas of respiratory bronchioles. Respiratory bronchioles in exposed animals contained aggregations of macrophages on their walls and within outpocketing alveoli (Figure 8). Increased numbers of cuboidal epithelial cells lined proximal respiratory bronchiolar surfaces, and there were decreased numbers of type I epithelial cells. Most of the cuboidal cells comprising this hyperplastic epithelial lining did not differ substantially from those in controls. Small numbers of cells contained apically placed, homogeneously electron-dense droplets, which appeared similar to the secretory granules found in nonciliated epithelial cells of terminal bronchioles.

Within the context of this chapter, the most important findings were that lesions result consistently in monkeys exposed to 0.2 ppm ozone, a

24

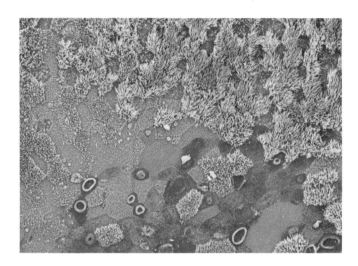

Figure 7. Scanning electron micrograph of trachea from a bonnet monkey exposed to 0.35 ppm ozone for seven days. There is marked shortening or absence of cilia from cells in the lower left portion of the field. Ciliated cells in the upper right portion of the field are relatively normal (X695).

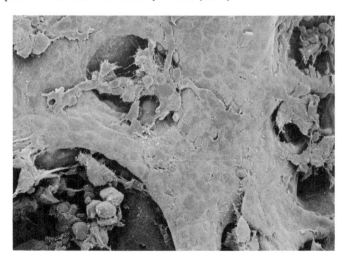

Figure 8. Scanning electron micrograph of proximal respiratory bronchiole from a bonnet monkey exposed to 0.2 ppm for seven days. Aggregations of macrophages are present in outpocketing alveoli. Respiratory bronchiolar wall is lined continuously by cuboidal cells (X416).

level of ozone commonly encountered by human populations, and that the site of the most severe damage is the respiratory bronchiole. In man, respiratory bronchiolitis has been identified as one of the most characteristic lesions in the peripheral airways of young cigarette smokers and has been postulated to be a precursor of centriacinar emphysema.[30] The results of our studies with monkeys suggest that oxidant air pollutants, such as ozone, might also result in damage to respiratory bronchioles of humans. The fact that most other laboratory animals do not have well-developed respiratory bronchioles points to the potential usefulness of primates in long-term studies designed to evaluate the contribution of air pollutants to the development of chronic obstructive airway disease. Bonnet monkeys and rats are being studied following a 90-day exposure to 0.8 ppm ozone. Preliminary results indicate that lesions in respiratory bronchioles persist in monkeys, whereas lesions in proximal regions of alveolar ducts diminish in severity with chronic exposure in rats.

Use of Primates in Sulfuric Acid Aerosol Studies

Monkeys have also been used in studies designed to assess the pulmonary effects of short-term exposure to sulfuric acid aerosol in the submicron droplet range.[31] Considerable species variations in response to exposure was observed among guinea pigs, mice, rats and rhesus monkeys. Whereas severe lesions characterized by necrosis of alveolar septa, alveolar edema, focal bronchiolar necrosis, and focal bronchial ulceration were present in guinea pigs exposed to 71 mg/m^3 (0.52 μm aerosol size) for four days, exposure to levels as high as 502 mg/m^3 (0.48 μm aerosol size) for seven days failed to produce lesions that were detectable by light microscopy and scanning electron microscopy. Exposure by others of cynomolgus monkeys to larger-sized droplets of sulfuric acid aerosol for periods of 78 weeks has revealed mild lesions in bronchi and respiratory bronchioles.[32,33]

OTHER FACTORS ASSOCIATED WITH THE USE OF MONKEYS AND DOGS

Although this chapter has emphasized the use of monkeys in studies on effects of air pollutants, this is not meant to diminish the importance of contributions made through the use of other animal species. Experiments with rodents offer the advantages of lower cost, relatively higher definition and control of age, weight and genetic diversity, as well as greater ease in handling and use of larger numbers in experimental studies. A major disadvantage of using these animals is that they do not have respiratory bronchioles. The laboratory dog is another animal that needs to be considered since it is large enough for detailed physiological studies and has well-developed respiratory bronchioles. They also are easier to obtain than monkeys and have an initial purchase price that is considerably lower (Table I). Differences in respiratory mechanics and distal airway structure between macaques and dogs have been discussed and the

Table I. Costs of Dogs and Monkeys in Air Pollution Studies

Item	Dog (pound)	Monkey (wild-caught)
Purchase	$ 27.80[a]	$ 225.00[a]
Quarantine/Condition	52.74	240.00
Cost Ready for Exposure	80.54	465.00
Per Diem (Average 7 California Vivaria)	1.99	1.53
Total 90-Day Exposure	259.64	593.70
Total 1-Year Exposure	807.10	1,014.45
Total 2-Year Exposure	1,533.15	1,572.90

[a]Costs are considerably higher when colony-bred animals are used.

closer similarity of these factors between macaques and man indicated. Dogs have an apparent advantage of being less expensive for use in short-term studies and when costs of experimental evaluation are low. When considering long-term studies where per diem costs exceed animal costs or studies with high costs of evaluation, however, economic advantages of using dogs rather than primates become negligible.

DISCUSSION AND CONCLUSIONS

This chapter has discussed the use of nonhuman primates in air pollution research and, specifically, their use in the evaluation of pulmonary effects of ozone and sulfuric acid aerosol. Comparative information on fine structure of distal airways and respiratory mechanics demonstrates advantages of using monkeys as an experimental model of man relative to other laboratory animals. An important finding from studies on effects of ambient levels of ozone on mankeys is that respiratory bronchioles are pulmonary structures that are most severely damaged. This is in contrast to findings from studies with rats. Rats do not have well-developed respiratory bronchioles, and studies with them have indicated that terminal bronchioles and proximal areas of alveolar ducts are the focus of the most severe damage.[10] Since man has well-developed respiratory bronchioles which appear to be sensitive to inhaled irritants,[30,34] studies on morphological or physiological effects of long-term exposure to air pollutants having known effects on distal airways would probably be better served if a species having respiratory bronchioles were used. Nonhuman primates are well suited for this role.

REFERENCES

1. Deinhardt, F., A. W. Holmes, R. B. Capps and H. Popper. "Studies on the Transmission of Human Viral Hepatitis to Marmoset Monkeys," *J. Exp. Med.* 125:673 (1967).
2. Gajdusek, D. C., C. J. Gibbs and M. Alpers. "Experimental Transmission of a Kuru-Like Syndrome to Chimpanzees," *Nature* 209: 794 (1966).
3. Hendrickx, A. G., L. R. Axelrod and L. D. Clayborn. "Thalidomide Syndrome in Baboons," *Nature* 210:958 (1966).
4. Wilson, J. G. "Thalidomide Syndrome," *Comp. Pathol. Bull.* 5:1 (1973).
5. Schmidt, L. H. "Selection of Species for Various Uses," in *Primates in Medicine*, Vol. 2, Using Primates in Medical Research, Part I. Husbandry and Technology, W. I. B. Beveridge, Ed. (New York: Karger, 1969).
6. Castleman, W. L., D. L. Dungworth and W. S. Tyler. "Cytochemically Detected Alterations of Lung Acid Phosphatase Reactivity following Ozone Exposure," *Lab. Invest.* 29:310 (1973).
7. Castleman, W. L., D. L. Dungworth and W. S. Tyler. "Histochemically Detected Enzymatic Alterations in Rat Lung Exposed to Ozone Ozone," *Exp. Mol. Pathol.* 19:402 (1973).
8. Stephens, R. J., M. F. Sloan, M. J. Evans and G. Freeman. "Early Responses of Lung to Low Levels of Ozone," *Am. J. Pathol.* 74: 31 (1974).
9. Stephens, R. J., M. F. Sloan, M. J. Evans and G. Freeman. "Alveolar Type 1 Cell Response to Exposure of 0.5 PPM O_3 for Short Periods," *Exp. Mol. Pathol.* 20:11 (1974).
10. Schwartz, L. W., D. L. Dungworth, M. G. Mustafa, B. K. Tarkington and W. S. Tyler. "Pulmonary Responses of Rats to Ambient Levels of Ozone," *Lab. Invest.* 34:565 (1976).
11. Evans, M. J., L. V. Johnson, R. J. Stephens and G. Freeman. "Renewal of the Terminal Bronchiolar Epithelium in the Rat following Exposure to NO_2 or O_3," *Lab. Invest.* 35:246 (1976).
12. Lum, H., L. W. Schwartz, D. L. Dungworth and W. S. Tyler. "A Comparative Study of Cell Renewal following Ozone or Oxygen Exposure. Response of Terminal Bronchiolar Epithelium in the Rat," *Am. Rev. Resp. Dis.* 115:228 (1977).
13. California Air Resources Board. "California Air Quality Data, July-September, 1974," VI, Sacramento (1974).
14. Bates, D. V., G. M. Bell, C. D. Burnham, M. Hazucha, J. Mantha, L. D. Pengelly and F. Silverman. "Short-Term Effects of Ozone on the Lung," *J. Appl. Physiol.* 32:176 (1972).

15. Hazucha, M., F. Silverman, C. Parent, S. Field and D. V. Bates. "Pulmonary Function in Man after Short-Term Exposure to Ozone," *Arch. Environ. Health* 27:183 (1973).
16. Hackney, J. D., W. S. Linn, D. C. Law, S. K. Karuza, H. Greenberg, R. Buckley and E. E. Pedersen. "Experimental Studies on Human Health Effects of Air Pollutants," *Arch. Environ. Health* 30:385 (1975).
17. Hackney, J. D., W. S. Linn, S. K. Karuza, R. D. Buckely, D. C. Law, D. V. Bates, M. Hazucha, L. D. Pengelly and F. Silverman. "Health Effects of Ozone Exposure in Canadians Versus Southern Californians," *Am. Rev. Resp. Dis.* 111:902 (1975).
18. Castleman, W. L., D. L. Dungworth and W. S. Tyler. "Intrapulmonary Airway Morphology in Three Species of Monkeys: A Correlated Scanning and Transmission Electron Microscopic Study," *Am J. Anat.* 142:107 (1975).
19. Von Hayek, H. "The Human Lung," translated by V. E. Krahl (New York: Hafner Publishing Co., 1960).
20. Watson, J. H. L., and G. L. Brinkman. "Electron Microscopy of the Epithelial Cells of Normal and Bronchitic Human Bronchus," *Am. Rev. Resp. Dis.* 90:851 (1964).
21. Andre-Bougaran, J., R. Pariente, M. Legrand and E. Cayrol. "Ultrastructure Normale des Petites Bronches et des Bronchioles Chez L'Homme," *Les Bronches* 24:1 (1974).
22. Kosch, P. C., J. R. Gillespie and J. D. Berry. "Respiratory Mechanics in Normal Bonnet and Rhesus Monkeys," *J. Appl. Physiol.* In press.
23. Kosch, P. C., J. R. Gillespie and J. D. Berry. "Flow-Volume Curves and Total Pulmonary Resistance in Normal Bonnet and Rhesus Monkeys," *J. Appl. Physiol.* In press.
24. Robinson, N. E., and J. R. Gillespie. "Lung Volumes in Aging Beagle Dogs," *J. Appl. Physiol.* 35:317 (1973).
25. Leith, D. E. "Comparative Mammalian Respiratory Mechanics," *The Physiologist* 19:485 (1976).
26. Grimby, G., and B. Söderholm. "Spirometric Studies in Normal Subjects. III. Static Lung Volumes and Maximum Voluntary Ventilation in Adults with a Note on Physical Fitness," *Acta Med. Scand.* 173:199 (1963).
27. Dungworth, D. L., W. L. Castleman, C. K. Chow, P. W. Mellick, M. G. Mustafa, B. Trakington and W. S. Tyler. "Effect of Ambient Levels of Ozone on Monkeys," *Fed. Proc.* 34:1670 (1975).
28. Mellick, P. W., D. L. Dungworth, L. W. Schwartz and W. S. Tyler. "Short Term Morphologic Effects of High Ambient Levels of Ozone on Lungs of Rhesus Monkeys," *Lab. Invest.* 36:82 (1977).
29. Castleman, W. L., W. S. Tyler and D. L. Dungworth. "Lesions in Respiratory Bronchioles and Conducting Airways of Monkeys Exposed to Ambient Levels of Ozone," *Exp. Mol. Pathol.* 26:384 (1977).
30. Niewoehner, D. E., J. Kleinerman and D. B. Rice. "Pathologic Changes in the Peripheral Airways of Young Cigarette Smokers," *New England J. Med.* 291:755 (1974).

31. Schwartz, L. W., P. F. Moore, D. P. Chang, B. K. Trakington, D. L. Dungworth and W. S. Tyler, "Short-Term Effects of Sulfuric Acid Aerosols on the Respiratory Tract. A Morphological Study in Guinea Pigs, Mice, Rats and Monkeys," Chapter 10, this volume.

32. Alarie, Y., W. M. Busey, A. A. Krumm and C. E. Ulrich. "Long-Term Continuous Exposure to Sulfuric Acid Mist in Cynomolgus Monkeys and Guinea Pigs," *Arch. Environ. Health* 27:16 (1973).

33. Alarie, Y. C., A. A. Krumm, W. M. Busey, C. E. Ulrich and R. J. Katz. "Long-Term Exposure to Sulfur Dioxide, Sulfuric Acid Mist, Fly Ash, and Their Mixtures. Results of Studies in Monkeys and Guinea Pigs," *Arch. Environ. Health* 30:254 (1975).

34. Heppleston, A. G. "The Pathological Anatomy of Simple Pneumo-koniosis in Coal Workers," *J. Pathol. Bacteriol.* 66:235 (1953).

CHAPTER 3

HEALTH EFFECTS OF AIR POLLUTION:
CONTROLLED STUDIES IN HUMANS

Jack D. Hackney, William S. Linn, Karl A. Bell
and Ramon D. Buckley

> Rancho Los Amigos Hospital Campus
> University of Southern California
> School of Medicine
> Downey, California 90242

INTRODUCTION

Controlled laboratory studies of humans exposed to air pollutants are relatively new. Before the 1970s, information on health hazards of inhaled substances came mostly from toxicological studies of laboratory animals or from epidemiological observations of people exposed to pollutants in the course of their normal activities. Although their scope is strictly limited by ethical considerations, controlled human studies have come to prominence in response to growing recognition of the need for better pollution control efforts guided by reliable scientific information on health hazards. Animal toxicology and epidemiology will continue to be important, but both fields have limitations which controlled human studies can help to overcome. Animal studies, if they employ genetically uniform animals in a rigorously controlled environment with careful quantification of pollutant dose and health effects, can provide the highest possible degree of scientific reliability (in the traditional sense). However, the relevance of animal toxic responses to human health risks is always more or less in doubt, since responses may differ in degree and kind between species. Furthermore, some important human health variables, particularly symptoms, have no readily observable counterpart in other species. The problems in epidemiology are quite the opposite of those in toxicology.

31

The relevance to human health is never at issue but the ability to control the experimental conditions (and thus to document toxic pollutant effects unequivocally) is almost always inadequate. Controlled human exposure studies can, in appropriate circumstances, combine some of the advantages of each of the more traditional investigative approaches and can thereby complement their findings to provide a greatly improved overall understanding of health consequences of air pollution.

EXPERIMENTAL PROCEDURES

Controlled human studies may be defined for present purposes as laboratory experiments in which volunteer subjects breathe atmospheres containing a single pollutant substance or a relatively simple combination of substances, under well documented and well controlled environmental conditions, realistically simulating real-world polluted environments of health interest. Subjects are tested for adverse health effects, during or after exposure, using the most sensitive means available. Satisfactorily designing and executing an investigation of this kind presents a number of complex problems and requires the participation of medical, biological and physical scientists. As in all human experimentation, the foremost concern must be to protect the safety and welfare of the subjects. This requires adequate screening of potential volunteers, obtaining informed consent of those selected for study, and careful monitoring for any untoward response during or after the exposure. The more directly scientific problems relate to generation of a well controlled, clean air background environment, choice of exposure conditions to simulate real-world exposures, generation and monitoring of the chosen pollutant atmospheres, and detection of exposure-related health effects.

Environmental Control and Monitoring

A necessary prerequisite for controlled studies is reliable knowledge of the nature of the pollutants in the atmosphere of health concern. Polluted air environments, community or industrial, usually contain complex and ever-changing mixtures of contaminants, not all of which can be monitored adequately. Usually, one or a few of the pollutants suspected of being most toxic on the basis of previous observations (most often by animal toxicology) are selected for controlled human studies. Ethically, the controlled pollutant exposure should never be more severe than might be encountered under "worst-case" conditions in the real world. Secondary stresses such as heat and exercise should be present in the experimental exposure if they typically accompany the real-world pollutant exposures being simulated. Obviously, the choice of exposure conditions should be based on careful review of extensive monitoring data from the environments of interest.

Some exposure studies have used breathing masks to deliver the test atmospheres, but this highly artificial exposure situation should be avoided if possible. Use of a controlled-environment chamber, large enough to permit relatively normal activity for subjects inside, is preferable. The more realistic the experimental exposures, the more meaningful are the results in relation to actual air pollution health effects. (One must keep in mind, however, that the response to a given pollutant substance in isolation may differ from the response to the same pollutant in the presence of many others which may also produce effects, independently or interactively. A properly cautious interpretation of findings would assume that additional pollutants most likely enhance adverse responses; thus, that responses to controlled single-pollutant exposures most likely underestimate the toxicity of real-world atmospheres.) Controlled-environment chambers are expensive to construct and operate, since air must be cleaned, conditioned and doped with pollutant substances on a fairly large scale. Elaborate monitoring equipment is also necessary to verify that air quality control is satisfactory. Few, if any, laboratories have the financial resources to employ "state-of-the-art" quality control technology. Fortunately, relatively simple purification systems may suffice in some investigations.

The air-cleaning technology used in biological and industrial clean rooms, employing high-efficiency particulate (HEPA) filters and laminar airflow schemes to prevent dissemination of any internally generated contaminants throughout the clean air area, is applicable to human exposure chambers. Additional capability is required to remove gaseous contaminants. Adsorption on activated carbon is an economical means of removing many pollutant gases, which, however, is not very effective against carbon monoxide, low-molecular-weight hydrocarbons, reactive gases like ammonia, or oxides of nitrogen. Permanganate-impregnated aluminum oxide pellets (Purafil Chemisorbant®) more effectively remove these low-molecular-weight compounds. Passing air successively through Purafil, carbon and HEPA filters provides relatively effective purification at a reasonable cost. Removal of carbon monoxide and additional hydrocarbons may be accomplished by oxidation to carbon dioxide on a high-temperature metal-oxide catalyst, *e.g.*, hopcalite or a precious-metal catalyst. This increases the cost greatly because of the expense of the catalytic converter itself and the extra air-conditioning capacity required downstream. A system equipped with a catalytic converter plus the above described filters (Table I) can deliver air nearly devoid of most pollutants. Methane and traces of other low-molecular-weight saturated hydrocarbons may remain but appear to be physiologically inert for most purposes and, thus, are not considered

Table I. Air Purification Materials Used For Controlled Environment Chamber Rancho Los Amigos Hospital, Downey, California

Material	Major Pollutants Removed	Principle
HEPA Filters	Particulates	Impaction
Activated Carbon	Large organic molecules, O_3	Adsorption, decomposition on surface
Purafil (H. E. Burroughs)	Oxides of nitrogen, other acid gases, readily oxidizable organics	Oxidation, acid neutralization
Hopcalite (MSA) High Temperature	CO, organics	Oxidation
Hopcalite (MSA) Ambient Temperature	Oxides of nitrogen	Adsorption, periodically heated to desorb

a problem. On the other hand, it must be kept in mind that trace contaminants in a controlled atmosphere, even if without direct effect on the experimental subjects, might react with experimentally generated pollutants to produce unsuspected (and perhaps biologically active) additional contaminants.

Gaseous pollutants are relatively simple to generate and monitor in exposure laboratories. For example, ozone, the most extenisvley studied pollutant gas, may be generated from air or oxygen using a high-voltage discharge or ultraviolet irradiation. It may be monitored by instruments based on colorimetry, coulometry, ultraviolet photometry or chemiluminescence. Since safety concerns are overriding, generated pollutants to which human subjects are exposed should be monitored by at least two independent instruments, preferably operating on different principles. Additional monitors for possible interfering pollutants, if available, will increase both the safety and scientific reliability of the experiment.

Particulate pollutants, particularly sufates produced by coal and oil combustion, are now receiving considerable attention as possible health hazards. Generation and monitoring of particulate pollutants for human studies is far more difficult than is the case with gases; thus, particulates have not yet been studied extensively. Principles and techniques from atmospheric chemistry, aerosol physics and chemical engineering must be employed. An initial problem is choosing appropriate aerosols to simulate ambient mixtures of interest (which are generally complex and variable in chemical composition, concentration and size distribution of the particles). One approach is to select an aerosol of one salt considered to be most representative. For example, ammonium sulfate is a logical choice to model ambient mixtures in which sulfate is the anion and ammonium the cation usually present in highest concentration. Size distribution and concentration of the experimental aerosol must be controlled well and carefully designed to simulate real-world conditions, since the size and concentration of the inhaled aerosol influences its transport properties, determining where in the respiratory tract it will be deposited (if at all) and, thus, what biological effects it may have. Since relative humidity and temperature affect particle size and transport properties, these parameters must also be controlled rigorously.

Several techniques are available for generating aerosols experimentally. Most are based on nebulizing an aqueous solution of the desired chemical compound into the test atmosphere. Monitoring of aerosols, either in the laboratory or in ambient air, is far more difficult than monitoring of gases, since techniques for continuous quantitative chemical monitoring in real time are generally not available. Chemical analyses usually require filter samples collected over periods of hours, and the sample composition may be subject to change between the times of collection and analysis. Continuous monitors are available to determine the number of particles in discrete size intervals per unit volume. In controlled exposures to a stable concentration of a single aerosol substance, count readings from such monitors may be related empirically to quantitative analyses of filter samples, thus providing an indirect means for continuous, quantitative aerosol monitoring.

Measuring Health Effects of Pollutants

Biological investigation in human exposure studies must aim at reliable detection of any health changes in exposed subjects. Further, it must attempt to differentiate between health changes resulting from the experimental pollutant challenge and those resulting from extraneous influences —a problem which may be thought of as separating signal from noise. The signal-to-noise ratio is often unfavorable in human studies because only relatively mild health responses may be studied ethically. Normal time variability of biological measurements or responses to a secondary stress may be mistaken for toxic responses to pollutants. Control experiments, consisting of exposures to purified air only, are always necessary to help differentiate pollutant effects from other biological effects. Ideally, control and exposure experiments should be done in random order, and the exposure conditions should not be known to the subjects or to the investigators immediately responsible for the health measurements. Usually, however, ethical and practical considerations require deviations from ideal design.

Exposure studies should simulate real exposures as closely as possible with respect to duration, temperature, humidity, exercise level and other relevant stress factors. Typically, exposures last a few hours per day for two or more days. Health-effect measurements must be designed to control for normal day-to-day and hour-to-hour variability to the extent possible. An important problem inherent in human studies is preventing interference from environmental stresses encountered by the subjects outside the immediate confines of the experiment. A common problem in our Los Angeles area laboratory is that if a pollution episode occurs on the day of a study, subjects may be exposed to nearly as high pollutant concentrations before or after the study as while in the exposure chamber. In principle, subjects could be kept in the chamber in purified air for long periods to minimize this interference, but such disruption of their normal activities might, in itself, constitute a psychological stress severe enough to alter some health-effect measures. Our approach is to attempt to schedule studies during times of low pollution levels, encourage subjects to remain indoors and at rest during intercurrent ambient pollution episodes, and provide them with portable air purifiers to be used in their homes or cars in some cases. Ambient air monitoring data for the period of a study are reviewed to give at least a rough idea of their intercurrent ambient exposures. Records are also kept of other obvious stresses that might affect experimental results. While the major purpose of such records is to safeguard against erroneous interpretations of the results, the data thus acquired may themselves be a useful object of investigation, since previously unrecognized stress-response relationships may be evidenced.

Inhaled pollutants first contact the tissues of the respiratory tract; then they may pass into the circulation, which transports them throughout the body. Many pollutants decompose before penetrating very far, bur their reaction products may still have relatively far-reaching effects. Respiratory function measurements are a logical first choice as an investigative tool in human exposure studies. Blood studies are also particularly valuable because blood, while less directly exposed than the respiratory system, can be sampled for direct examination, unlike other internal tissues. Symptoms experienced by the subjects during exposure, particularly those related to breathing, are another important target of investigation. While symptoms are subjective and hard to quantitate, they are intimately related to the ability to perform normal tasks—an especially important aspect of health.

Respiratory function measurements are designed to test for alterations in lung mechanics, i.e., disturbances of the normal pressure-flow-volume relationships or of the normal distribution of air to the various regions of the lung. Changes in maximum attainable air flowrates and resistance of the airways are given considerable attention because they are easy to measure. Measurable function changes under pollutant exposure may occur because irritation of the airways makes maximal breathing efforts uncomfortable for the subject, because exposure induces bronchial constriction (as in an asthmatic attack) or because the alveolar-capillary (blood-air) interface is disturbed, *e.g.*, by an influx of fluid into the airspaces or by an alteration of the surfactant material lining these spaces. Determining the exact cause of a given observed function change is often difficult.

Blood studies in humans focus on enzyme activity levels, reactions of lipids with pollutant molecules and structural integrity of cells. Experimental approaches are commonly derived from those used previously in animals. Much work has been done with animals and oxidant pollutants, and this has been reviewed recently.[1] Carbon-carbon double bonds (abundant in lipids) and sulfhydryl groups (abundant in proteins) have been found to be highly susceptible to oxidant attack. Thus, assays for essential substances in human red cells which contain these functional groups are of interest. Biochemical changes in humans exposed to pollutants can be detected readily; however, other stresses affect blood biochemistry also, making careful control experiments especially important.

SOME RECENT FINDINGS

Space does not permit even a brief review of the entire field of controlled human studies, so this report is limited to recent work in our laboratories and others closely related (Table II). The emphasis is on pollutants typically found in photochemical smog of the type long associated with Los Angeles but now recognized in many other areas.

Ozone

Ozone (O_3) is the most powerful oxidizing agent among common pollutant gases and is known to be highly toxic from animal and occupational health studies. Maximum atmospheric concentrations during smog episodes sometimes exceed 0.6 part per million (ppm), averaged over one hour, in the Los Angeles area. Bates and co-workers[2-4] first reported ozone studies in humans using an exposure chamber and a protocol designed to simulate ambient exposures. They studied healthy Canadian volunteers, not frequently exposed to photochemical smog, breathing 0.75 or 0.37 ppm O_3 for two hours while intermittently resting and exercising lightly on a stationary bicycle. The total amount of exercise was similar to that which might be experienced while performing light outdoor work for a similar period. Respiratory symptoms and decrements in lung function

Table II. Pulmonary Effects Observed in Some Recent "Realistic" Controlled Exposures to Common Pollutant Gases[a]

Gas	Concentration (ppm)	Response	Remarks
O_3	0.37	+	Healthy Canadians[3,4,11]
O_3	0.37	+	Healthy new arrivals to Los Angeles[12]
O_3	0.37	-	Healthy Los Angeles residents[11,12]
O_3	0.50	+	Los Angeles residents; response most common in those with history of respiratory hypersensitivity[7,8,13]
$O_3 + SO_2$	0.37 + 0.37	+	Healthy Canadians; sulfates probably also present[18,19]
$O_3 + SO_2$	0.37 + 0.37	±	Los Angeles residents; little or no sulfate[19]
$O_3 + NO_2$	0.50 + 0.30 or 0.25 + 0.30	-	Los Angeles residents not reactive to O_3 alone at same concentration[7,8]
SO_2	0.37	-	Canadians, Los Angeles residents[18,19]
NO_2	1.0	-	Healthy Los Angeles residents[14,15]

[a]Exposures in a chamber for $\geq$ 2 hours with intermittent exercise.
[b]"+" = definitie loss in lung function test performance and respiratory symptoms in some subjects: "-" = no such response observed; "±" = slight or equivocal response.

test performance were found at both concentrations and were dose-related (*i.e.*, more severe at higher doses). In our laboratory, Los Angeles residents were studied using a similar experimental protocol with the addition of elevated temperature (31°C)—a secondary stress usually present during Los Angeles smog episodes. Respiratory symptoms, function changes and blood biochemical changes were observed at 0.37 and 0.50 ppm O_3, but not at 0.25 ppm.[5-8] Individual responses varied considerably. Some subjects tolerated 0.50 ppm for as long as five hours without symptoms or function changes, while others reported symptoms after as little as one and one-half hours. Typical responses included cough, soreness in the throat and chest, small losses in forced expiratory flowrates and vital capacity (the maximum volume of air one can expire), slightly impaired distribution of ventilation within the lung, increased fragility of red blood cells and slight losses in red cell enzyme activities. Symptoms and lung function changes tended to occur together but seemed unrelated to blood changes. All three types of response appeared to increase with O_3 dose.

Since some, but not all, people suffer obvious adverse effects of O_3 at dose levels attainable in ambient smog exposures, an important scientific public health question arises: What factors predispose individuals to adverse reactions? One probable risk factor is preexisting respiratory disease. Among our Los Angeles subjects, those who denied any history of respiratory problems seldom reacted measurably in two-hour exposures to 0.50 ppm, while those with respiratory allergies or mild asthma usually did react under the same conditions. Moderately asthmatic subjects sometimes showed marked reactions at 0.37 ppm. While a rigorous statistical comparison has not been made, it appears highly likely that asthmatics are at increased risk of adverse reactions to O_3, based on our overall findings in Los Angeles residents. When these results are compared with the previous Canadian studies,[3,4] however, an interesting discrepancy is evident. Healthy Canadian subjects appear to have reacted far more severely than healthy Los Angeles residents. While this difference might have been related to experimental methodology, it seemed likely that Los Angeles residents might have "adapted" to O_3 exposure since they, unlike the Canadians, frequently breathed photochemical smog. Many laboratory animal studies support this possibility.[9,10] Accordingly, we undertook a comparative study in cooperation with the Canadian investigators[11] and a similar study of a preprofessional student group.[12] In both cases, with subject groups being matched as closely as practical for other factors, non-Los Angeles residents showed more symptoms and larger mean lung function changes than did Los Angeles residents in two-hour exposures to 0.4 ppm. In an attempt to demonstrate O_3 adaptation more directly, we exposed a group of Los Angeles residents with respiratory allergy or mild asthma to 0.5 ppm, two hours per day on four successive days. This was done during the winter (low-smog) season to minimize the chance of preexisting adaptation. Symptoms and function changes occurred with the first two O_3 exposures but were largely absent in the last two exposures.[13] Thus a good deal of evidence supports the possibility of adaptation in humans. The underlying biological mechanism of the apparent adaptive response remains unknown. Another important unanswered question concerns the possibility of as yet undetected, harmful long-term consequences.

Nitrogen Dioxide

Nitrogen dioxide (NO_2) has a much lower oxidation potential than O_3 and is less toxic at similar concentrations. One-hour average concentrations in Los Angeles ambient smog have reached as high as 1.0 ppm. We found no substantial lung function or symptom changes in healthy Los Angeles residents exposed to that concentration for two or three hours under conditions like those in the previously described O_3 studies.[14,15] In a limited number of studies simulating "worst-case" pollution episodes with NO_2

and O_3 coexisting, no enhancement of toxicity relative to O_3 alone was observed.[7,8]

Sulfur Oxides

Sulfur dioxide (SO_2) and particulate sulfate pollution result primarily from fossil fuel combustion; however, SO_2 may be more effectively converted to sulfate particles in photochemical smog. Ambient SO_2 and sulfate aerosol concentrations may exceed 1.0 ppm and 100 $\mu g/m^3$, respectively, in severely polluted areas. Controlled human studies generally have shown measurable adverse effects of SO_2 only at higher-than-ambient concentrations, but many epidemiological studies[16,17] have suggested associations between ambient SO_2 or sulfate levels and illness and death rates. Hazucha and Bates[18] found that a group of healthy Canadian subjects exposed for two hours to SO_2 and O_3 together, each at 0.37 ppm, showed much more severe lung function changes and symptoms than similar groups exposed similarly to O_3 alone (which produced less severe changes) or SO_2 alone (which produced no measurable changes). A similar study of Los Angeles subjects in our laboratory showed only slightly greater effects of the mixed gases than of O_3 alone.[19] The test gases were found to react in the environmental chamber system to produce sulfur-containing particles, probably sulfuric acid or a sulfate salt. These aerosol products may be more important in toxic responses than SO_2 itself, both in laboratory studies and in ambient exposures. Controlled human studies of aerosols of individual sulfate salts are in progress in our laboratory and others; so far, no adverse short-term effects have been reported at concentrations likely to be attained in ambient air.

Many questions remain to be answered in the area of sulfur-oxide toxicity, which is of particular concern in light of the increasing reliance on high-sulfur fuels. The O_3-sulfur-oxide interaction is still not well understood, and many other pollutant interactions of possible health significance have not been investigated at all. Physical and chemical behavior of pollutant substances both in the atmosphere and in living organisms need to be far better understood for the health consequences of air pollution to be evaluated realistically.

DISCUSSION AND CONCLUSIONS

Controlled human studies can and should provide a major part of the scientific guidance for pollution control planning and air quality standard setting. We have described some of the recent progress in this field as well as some of the problems in experimentation and interpretation of results that may be expected. Considerable attention has been given to experimental methodology and to interdisciplinary efforts, because these factors are crucial to success. Contributions from chemistry and physics are needed not only to provide satisfactory atmospheric monitoring and control, but also to help explain the ways in which pollutants interact with each other and with biological molecules. Biology can further examine the macroscopic as well as the molecular effects of pollutants on living organisms. In animal biology, particular emphasis should be placed on long-term effects of repeated exposures and on the relationships between short- and long-term effects. Human studies are ethically limited to short-term reversible toxic effects; thus, the judgement whether short-term effects are linked to long-term irreversible damage must be based on observations in animals. Short-term responses alone are sufficient to demonstrate that ozone may harm humans at elevated ambient concentrations. The existing evidence is less conclusive for other common pollutants, but there are suggestions of interactive effects rendering mixtures more harmful than single substances. Continued vigorous effort is needed to identify and understand air pollution health hazards.

REFERENCES

1. Cross, C. E., A. J. DeLucia, A. K. Reddy, M. Hussain, C. K. Chow and M. G. Mustafa. "Ozone Interactions with Lung Tissue," *Am. J. Med.* 60:929 (1976).
2. Bates, D. V., G. Bell, C. Burnham, M. Hazucha, J. Mantha, L. D. Pengelly and F. Silverman. "Problems in Studies of Human Exposure to Air Pollutants," *Can. Med. Assoc. J.* 103:833 (1970).
3. Bates, D. V., G. Bell, C. Burnham, M. Hazucha, J. Mantha, L. D. Pengelly and F. Silverman. "Short-Term Effects of Ozone on the Lung," *J. Appl. Physiol.* 32:176 (1972).
4. Hazucha, M., F. Silverman, C. Parent, S. Field and D. V. Bates. "Pulmonary Function in Man After Short-term Exposure to Ozone," *Arch. Environ. Health* 27:183 (1973).
5. Buckley, R. D. J. D. Hackney, K. Clark and C. Posin. "Ozone and Human Blood," *Arch. Environ. Health* 30:40 (1975).
6. Hackney, J. D., W. Linn, R. Buckley, E. Pedersen, S. Karuza, D. Law and A. Fischer. "Experimental Studies on Health Effects of Air Pollutants. I. Design Considerations," *Arch. Environ. Health* 30:373 (1975).
7. Hackney, J. D., W. Linn, J. Mohler, E. Pedersen, P. Breisacher, and A Russo. "Experimental Studies on Human Health Effects of Air Pollutants. II. Four-hour Exposure to Ozone Alone and in Combination with Other Pollutant Gases," *Arch. Environ. Health* 30:379 (1975).
8. Hackney, J. D., W. Linn, D. Law, S. Karuza, H. Greenberg, R. Buckley and E. Pedersen. "Experimental Studies on Human Health

Effects of Air Pollutants. III. Two-hour Exposure to Ozone Alone and in Combination with Other Pollutant Gases," *Arch. Environ. Health* 30:385 (1975).

9. Morrow, P. E. "Adaptations of the Respiratory Tract to Air Pollutants," *Arch. Environ. Health* 14:127 (1967).

10. Fairchild, E. J. "Tolerance Mechanisms. Determinants of Lung Responses to Injurious Agents," *Arch. Environ. Health* 14:111 (1967).

11. Hackney, J. D., W. S. Linn, S. K. Karuza, R. D. Buckley, D. C. Law, D. V. Bates, M. Hazucha, L. D. Pengelly and F. Silverman. "Effects of Ozone Exposure in Canadians and Southern Californians: Evidence for Adaptation?" *Arch. Environ. Health* 32:110 (1977).

12. Hackney, J. D., W. S. Linn, R. D. Buckley and H. J. Hislop. "Studies in Adaption to Ambient Oxidant Air Pollution: Effects of Ozone Exposure in Los Angeles Residents vs. New Arrivals," *Environ. Health Pers.* 18:141 (1976).

13. Hackney, J. D., W. S. Linn, J. G. Mohler and C. R. Collier. "Adaptation to Short-Term Respiratory Effects of Ozone in Men Exposed Repeatedly," *J. Appl. Physiol.* 43:82 (1977).

14. Hackney, J. D., F. C. Thiede, W. S. Linn and E. E. Pedersen. "Effect of Short Term NO_2 Exposure on Lung Function in "Normal" Human Subjects," (Abst) *Chest* 64(3):395 (1973).

15. Hackney, J. D., F. C. Thiede, W. S. Linn, E. E. Pedersen, C. E. Spier, D. C. Law and D. A. Fischer. "Experimental Studies on Human Health Effects of Air Pollutants. IV. Short-Term Physiological and Clinical Effects of Nitrogen Dioxide Exposure," *Arch. Environ. Health.* 33:176 (1978).

16. National Air Pollution Control Administration. "Air Quality Criteria for Sulfur Oxides," Pub. No. AP-50 (Washington, DC: U.S. Government Printing Office, 1969).

17. Environmental Protection Agency, U.S. "Health Consequences of Sulfur Oxides: A Report from CHESS, 1970-1971," Publication No. EPA/650/1-74-004 (Washington, DC: U.S. Government Printing Office, 1974).

18. Hazucha, M., and D. V. Bates. "Combined Effect of Ozone and Sulfur Dioxide on Human Pulmonary Function," *Nature* 257:50 (1975).

19. Bell, K. A., W. S. Linn, M. Hazucha, J. D. Hackney and D. V. Bates. "Respiratory Effects of Exposure to Ozone plus Sulfur Dioxide in Eastern Canadians vs Southern Californians," *Am. Ind. Hyg. Assoc. J.* 38:696 (1977).

Ref.

CHAPTER 4

RELATIONSHIP OF LONG-TERM
ANIMAL STUDIES TO HUMAN DISEASE

Jerry F. Stara and Dinko Kello

Environmental Criteria and Assessment Office
U.S. Environmental Protection Agency
Cincinnati, Ohio 45268

INTRODUCTION

In the past, some rulers used their subjects as human tasters to assure the safety of their food.[1] Results of human experience were then and are still the primary basis for accepting our diet. Quoting L. B. Jensen's book, *Man's Foods*, Friedman[1] tells how primitive men fed their dogs new fruit, vegetables or any other unknown food to see whether the food was poisonous. Only if the dogs did not become ill did they try it themselves. This is probably how the science of toxicology began many centuries ago.

Although predictive tests in toxicology have now increased in number and reached a high level of sophistication, "safe" doses still cannot be established with certainty, even with the increasingly complex battery of tests. We have increased the number of mice, rats and dogs, added species such as primates, marmosets, pigeons, quail, miniature pigs and goats, and increased the longevity of tests from ten days, to two weeks, to three months, to two years and even to a life span or to three generations. Initially we were concerned with only toxicity; more recently we have begun to test for carcinogenesis, mutagenesis and teratogenesis using multigeneration tests in several species and strains. Animals were outbred and inbred, randomly and nonrandomly mated; however, none of this significantly improved the prediction process for man. Many years ago we counted dead rats or mice, but this, of course, was incorrect. Now we rely on tests with blood and urine, and every organ that can be removed is sectioned and examined histologically. We look at every functional change possible. We stepped down from cellular to subcellular level, added phar-macokinetic, biochemical and behavioral studies, radioactive tracers and electron microscopy. The vast amount of duplication and duplicity in scientific experiments with rats and mice and the application of the data to human populations have become a joking matter to the layman. More and more scientists in various disciplines are attempting to predict human risk by the many thousands of physical and biological agents that have been introduced into human environment. However, we still follow the "golden rule" in toxicological evaluations of these environmental pollutants, to ascertain their dose-effect relationships, to set minimal toxicity levels or thresholds and, as a consequence, to establish some permissible limits.[2]

In an attempt to deal with the overwhelming number of chemicals, the U.S. government has promulgated various legislative acts aimed at controlling the proliferation of potentially harmful compounds and at protecting pub-lic health. Some 12 government agencies are involved in various aspects of monitoring, evaluating, regulating or enforcing these legislative mandates. For example, the FDA has a major responsibility for controlling drugs, food and food additives; OSHA enforces the control of occupational exposure of workers; and EPA is responsible for controlling toxic agents in the environment (air and water). The Toxic Substances Control Act (TSCA) of October 11, 1976[3] may prove to have the widest influence on environmental pollution control and abatement.[4-7] The full effect of this new act has not yet been fully realized. As it is translated into specific regulatory and compliance requirements, the act will probably be interpreted and reviewed by all parties, including the chemical industry and the smaller manufacturers, processors, formulators and users. Eventu-ally, the act should provide a complete guide for the EPA and various affected companies.

For the regulatory measures to be successful, there are major issues that must be solved by the scientific community. Falk[5] sums up the most important problems concerning the need for testing the overwhelming number of the potentially hazardous substances: "Many toxic substances reach the environment which have not been tested for their toxicity whatsoever; and if we assume that they all must be tested, there may not be enough mice or rats or even investigators around to do the job." The amount of needed work is staggering and related cost is prohibitive. Further, Falk states, "the question of which tests are necessary for a *conclusive* (satisfactory) answer on product safety for mankind cannot be answered with present knowledge and consequently, should not even be asked."

Recently a number of investigators considered various aspects of this program. This chapter reviews the present dilemma regarding testing of chemicals and the major factors contributing to it, and makes some suggestions on how to deal with this monumental task. Particular emphasis is given to selecting appropriate requirements for chronic toxicological studies in mammals. The need for a carefully selective approach is stressed in planning long-term, low-level studies and special attention is given to studies in higher mammalian species of chronic diseases in man.

THE ENVIRONMENTAL DILEMMA

Many factors, both technical and practical, beset a toxicologist trying to evaluate human risk accurately from the myriad of potentially toxic agents. Major problems must be distinguished from minor issues, which can be dealt with properly if good judgment is exercised. It is the overall factors that must be resolved by scientists and regulators together. The four major issues are suggested to be *(1) establishing the priority of chemicals for testing; (2) cost; (3) dose; and (4) extrapolation of experimental animal data to man.*

Scientific research has identified more than 4.3 million chemical compounds, some 63,000 of which are thought to be in common use in this country. This number grows almost 10% each year.[8] It is estimated that perhaps 1,000 new compounds a year may enter the market worldwide.[6] In the U.S., the number of new compounds entering the market each year ranges between 300-700.[8,9] In determining which environmental agents should be placed on top of the evaluation list, there are several possible approaches, making this task extremely difficult.

Certainly two major factors are the *expenditures and resources* required to conduct an acceptable testing program. The present cost of a comprehensive, extensive toxicological test battery to assess satisfactorily human risk (including carcinogenic, teratogenic, mutagenic) may run as high as $800,000 for one chemical. A less complete set of tests may cost $200,000-300,000, with an average for a presently acceptable standard testing protocol of about $500,000 per compound.[10] Additional cost includes environmental assessment studies such as the ecological tests on aquatic, marine and freshwater organisms, plants, meat- and milk-producing animals, and wildlife. The present toxicological and epidemiological research program supported by the U.S. government is estimated at $100 million annually which, if computed on a cost per compound basis, may provide a complete set of data for 200-300 compounds a year. However, a number of these studies are duplicates (in some cases this is necessary), so the number of compounds actually tested is probably significantly lower. Expenditures by private industry are not known at present.[4]

In 1538, Paracelsus stated that "all things are poison and none are not and that only *the dose* determines that a thing is not a poison." Most modern toxicologists still agree, certainly for noncarcinogenic substances. Dose alone makes a chemical poisonous; within practical limits a no-effect level can be determined in most animal experiments, even though statistical considerations may indicate that the limited number of animals cannot prove the point definitely.

Low-level occupational or environmental exposure to a potentially carcinogenic chemical creates a unique problem particularly for the regulatory agencies because they must decide at what level the exposure should be controlled. Controversies have arisen because of the nature and inadequate test regime for long-term effects. Since results of animal carcinogenic studies usually are obtained using high doses of the compound in relatively small numbers of animals, the resulting extrapolation to large human populations who are exposed to much *smaller* amounts is extremely difficult, if not impossible.

The issue that justifiably continues to receive a great deal of attention and which was often thought to be insurmountable is *extrapolation* of animal data to man. Alexander Pope said, "The proper study of mankind is man." However, man cannot be used as an experimental animal for obvious sociological/medical/legal reasons. Lower animals must be used to estimate the risk for humans. Webster defines *extrapolation* as "to project, extend, or expand known data or experience into an area not known or experienced so as to arrive at a usually conjectured knowledge of the unknown area by inferences based on an assumed continuity, correspondence, or other parallelism between it and what is known." Estimating risk for an unknown compound presently includes initial screening procedures, *i.e., in vitro* bioassay, which is usually accompanied by acute animal toxicity tests. This procedure is sometimes followed by a subacute and a chronic toxicity evaluation of the most suspicious candidate-agents. The inference is probably more tenuous when attempts are made to extrapolate directly from the cellular and molecular level to man. The inferences made from the *in vivo* experimental animal data may be less tenuous but they still exist. However, since to perform a great majority of these experiments on human subjects is unacceptable, we must accept a certain degree of uncertainty in extrapolating results from animals. Zapp[10] says that "we must always keep a measure of skepticism about any extrapolation until it has been verified in human experience," which, of course, is out of the question.

Figure 1 illustrates the most essential problem in estimating human risk—trying to determine what happens at lower exposures from animal bioassays usually obtained at relatively high levels of exposure (as indicated by the triangles). Extrapolation of the observed results with a straight

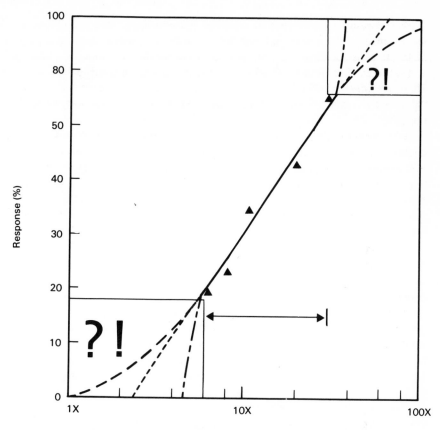

Figure 1. The problems in estimating human risks from animal bioassays.[11]

line yields a result suggesting no response at low doses, *i.e.,* a threshold. Scientists continue to disagree whether the chemical carcinogens are metabolized similarly to other toxic chemicals or through a different pathway resulting in the unrestrained growth of cells and irreversibility; also, whether the low levels of chemical carcinogens have the capacity to induce tumorigenesis or whether they behave similarly to other chemicals that appear to have a threshold at low doses, indicating a detoxifying mechanism. Under properly designed experimental conditions the threshold hypothesis assumes that a no-effect dose for a chemical can be determined, even though the concept of a "no-effect" dose seems to be losing popularity. If this chemical is a carcinogen, a no-effect dose implies that induction of tumors either cannot occur or is extremely improbable. Scientists who subscribe to this theory believe that the low "environmental" levels of chemicals do not present a significant carcinogenic hazard because of their measured occurrence at very low concentrations in the environmental media. The linear, or "one-hit" hypothesis assumes that every molecule of a chemical carcinogen reacts with a DNA receptor in a cell of the body. The dose-response curve is linear, going through zero dose. There are other models being considered, assuming that two or more "hits" are necessary for cancer induction, but all theorize the lack of a threshold.

As a result of this controversy, regulatory agencies face a serious dilemma which must be solved soon if they are to regulate environmental and occupational exposures to chemicals (including carcinogens) at defined levels in food, air and water, even when data are practically nonexistent. There is limited evidence for both points of view; however, regulatory agencies lean toward the more conservative view that carcinogens do not exhibit thresholds. Examples of this conservative approach include the "Delaney Amendment" to the Food and Drug Act, which prohibits additions to foods of chemicals that have been shown to be carcinogenic in animals or man; the Federal Insecticide, Fungicide and Rodenticide Act, similarly prohibiting the use of pesticides that have been shown to be carcinogenic; and proposed regulation by OSHA, which would prohibit industrial use of carcinogens, unless there is no alternative (and then limited exposure would be permitted only under strictly controlled conditions[7]). As stated before, scientists believe both that chemicals causing cancer at high levels can be detoxified at very low levels and, vice versa, but neither side can provide sufficient data to prove its case. Maugh[7] lists evidence both for and against the two hypotheses. For example:

1. Scientists at the Franklin Institute found that the dose-response curves for 151 chemicals studied in experimental animals supported the no-threshold hypothesis.
2. The National Academy of Sciences stated in several reports, including the 1977 Report on Drinking Water and Health,[12] that no substantial evidence exists for the existence of a threshold and, therefore, in the absense of definitive data it must be assumed (and perhaps erred on the safe side) that thresholds do not exist.
3. Dr. Gehring of Dow Chemical Co., as an industry representative, disagrees and states that the no-threshold approach doesn't follow common sense,[11] and suggests that up to 95% of the chemicals in existence can react with DNA. Perhaps half of them are already in a reactive form and the others can be transformed into metabolites by various enzymes systems. The defense mechanisms of the body include cellular membranes, which prevent many chemicals from coming into contact with DNA, the enzyme systems that repair damaged DNA, the enzyme systems that detoxify foreign substances, and the immune system which may destroy cells damaged by foreign substances.

Magee[13] recently summarized the findings on the rates of removal of different alkylated bases from rat DNA treated with nitrosoalkylating agents *in vivo*. The data strongly indicate that many DNA repair processes occur at the molecular level in the intact animal.[14-16] These findings have obvious implications for the existence of defense mechanisms in the body against exposure to very low levels of chemical carcinogens and mutagens. The best evidence supporting the dose-response linear or one "hit" theory for carcinogenesis is the radiation-induced leukemia data which follow the linear dose response curve down to an induced incidence of 0.1%. However, when some scientists insist that chemicals would follow the same dose response curve in production of cancer, others disagree and point out that radiation as a physical entity interacts with DNA by producing free radicals in the inter- and intracellular region at random. A chemical substance, on the other hand, must follow an established metabolic pathway and be transported by the use of several mechanisms of transport before it can react with DNA, and, therefore, must cope with many obstacles before it can reach the nucleus of a cell. Falk[5] explained this in some detail when he stated that many chemicals, carcinogenic or not, will interact with intracellular elements other than DNA or may interact with DNA without initiating a tumor; only an interaction with a few specific sites of DNA tends to induce tumor formation. Thus, the probability of a chemical carcinogen reacting at the exact site should be extremely low for very small doses of chemicals. The fact that cellular repair mechanisms exist protecting the DNA and that the immune system often destroys damaged cells give further evidence as to the protective mechanisms at low levels. Furthermore, the relationship between dose of chemical carcinogens is generally recognized to be inversely proportional to the latent period between initiation of exposure and tumor appearance. Jones and Grendon[17] implied from their calculations that this relationship presents evidence of a "practical" threshold, since at very low doses the latent period may amount to several times the animal life span. However, the suggestion of a "practical" threshold also has been severely criticized and the more conservative "no-threshold" position has generally been maintained.

An *ad hoc* committee commissioned by the Surgeon General to consider possible changes in the Delaney clause reported that "The principle of zero tolerance for carcinogenic exposures should be retained in all areas of legislation presently covered by (the Delaney clause) and should be extended to cover most (other) exposures as well. . . . Exceptions should be made only after the most extraordinary justification."[7] In spite of these controversies, the toxicology profession can be proud of the progress made during the past half century in discovering that many chemical and physical agents play a significant role in the etiology of cancer.

Barry Commoner[18] summarized the progress of carcinogenic studies in experimental animals from 1930 to the present. Figure 2 shows a rapid

50

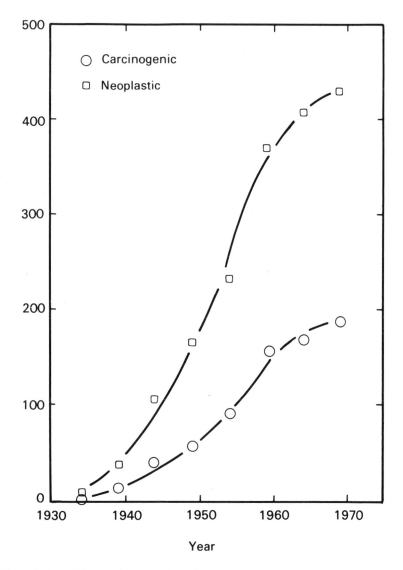

Figure 2. Annual increase in the number of carcinogenic substances discovered by tests with animals.[18]

increase in identification of tumorigenic and carcinogenic substances in laboratory animals during this period. These results underline the generally accepted assumption that 75-85% of all cancers are due to toxic agents found in the environment, and that perhaps 10-15% of *all* environmental pollutants may be carcinogenic in nature.[8,19,20]

Yet, the list of proved or suspected human carcinogens is very small (Tables I, II), even though tested rats or mice "do not smoke, do not breathe hydrocarbons or sulfur oxides from fossil fuels, do not take drugs, do not drink alcohol, and do not eat bacon or smoked salmon or well-done hamburgers."[21]

Table I. Recognized Human Carcinogens[22]

Anthracene Oil	Isopropyl Oil
Arsenic	Lignite Oils
Asphalt	Mustard Gas
Asbestos	Naphtylamine
Benzidine	Nickel
Benzol	Paraffin Oils
Carbon Blacks and Soots	Pitch
Chlonarchis Sinensis	Radioactive Nuclides
Chlorobenzols	Schistosoma haematobium
Coal Tar	Shale Oil
Creosote	Ultraviolet Irradiation
Chromium	X-irradiation
Fuel and Lubricating Oils	Xenylamine

Table II. Probable Human Carcinogens[13]

Compound	Cancer
Chlornaphazine (an aromatic amine derivative used as a drug)[23]	Carcinoma of the bladder
Diethylstilbestrol[24]	Carcinoma of the vagina
Anabolic steriods[25]	Carcinoma of the liver
Contraceptive steriods[26]	Benign hepatoma
Vinyl Chloride[27]	Hemangiosarcoma of the liver
Bis(chloromethyl) ether[28]	Carcinoma of the lung

Zapp,[10] Safiotti[20] and others conclude there is no mathematical method that can extrapolate data from any animal experiment to give a guaranteed human safe level (and that is certainly true of a so-called safe level for a carcinogen), but a realistic "no-effect" level of exposure to the overwhelming majority of environmental agents may exist. We cannot scientifically establish an absolute safe level, even from epidemiological studies of workers, but in a number of instances the work experience and industrial health data imply that compounds shown to be carcinogenic in rats have not demonstrated increased incidence of cancer in workers during their lifetime. Most scientists would agree that biological effects observed in experimental animals cannot provide an absolute proof of the safety of a chemical agent in man, but this is also true of human data since clinical experience in a few members of human species cannot be extrapolated to the general U.S. population. Ultimate data on human effects can be obtained only by recording disease incidence over a long period of time in large population groups. Consequently, only much more sophisticated and effective surveillance of the health of the population will enable detection of low-level effects in man, rather than extrapolation from the strikingly acute effects that we are able to observe at present.

Another factor seriously affecting risk assessment is that the results of a full testing program are not available until two to ten years after initiation of the studies, depending on the test array and the species used. This long lag time increases the already pessimistic picture. As a result, a large number of chemicals are being used, particularly in the industrially advanced countries, but no evaluation criteria are being put into effect to protect the consumer. In addition, some of the tests that were performed (primarily those using mice) are now believed to have provided insufficient data for satisfactory extrapolation to man. This creates another dilemma for the science of toxicology: Should studies on many "old" chemicals be repeated using the presently recognized standard test for carcinogenic assessment of "new" chemicals (which, according to TSCA, includes the oral administration of the compound to rats with a second test in another rodent species) in order to provide data that are suitable for extrapolation?

Many other factors contribute to the dilemma of how to proceed quickly, economically and on a sound scientific basis. Some factors are directly related to experimental design and will be reviewed later in this chapter. Yet dominating all others is the need for judicious and scientific prioritization of environmental chemicals for short-term screening followed by intermediate and chronic toxicological assessment.

KEY APPROACHES TO TOXICOLOGICAL
EVALUATION OF ENVIRONMENTAL POLLUTANTS

The changing approach in the evaluation of chemical substances has elicited a number of reports on the subject.[10,13,29-32] The major issue in all reports is the best approach for choosing the compounds to be studied and the appropriate methodology to obtain satisfactory data. The array of tests in experimental protocols is continuously being revised and updated across the whole area of short-term bioassays and *in vivo* testing procedures. Acute and subchronic toxicity protocols are being supplemented by other tests to evaluate the potential mutagenic, teratogenic, carcinogenic and other long-term effects of all compounds being or to be marketed.

A number of highly respected scientists have discussed in depth the over-all approach to evaluating the effects of chemical compounds on the popula-tion from the many options available. Every study pertaining to the evaluation of low levels of toxic agents and their effect on public health is fallible, but an orderly working plan must be developed and a forceful program must be initiated despite inadequacies that are often based on technical, political or personal considerations.

Different approaches to the identification of environmental pollutants hazardous to human health have been proposed. Under the Guidance of National Science Foundation, a group of 10 prominent members of the United States scientific community was assembled in February, 1974 to review data on production, use, disposal, properties and toxicity of certain organic com-pounds in commercial production. The primary goal was to identify and select from the manufactured organic chemicals those compounds that may be of present or future interest with respect to their possible environmental or health effects. A final list of 80 compounds, popularly known as the "Most Wanted Criminal List," was derived by six basic criteria suggested previously by Korte:[31]

1. their production and industrial waste;
2. the use pattern;
3. persistence;
4. the dispersion tendency;
5. conversion under biotic and abiotic conditions; and
6. the biological consequences.

Table III shows the top 10 in the priority list of 80 compounds considered hazardous to the environment and human health.

Following the investigation of health hazards in the chemical industry, Selikoff[30] outlined the criteria for determining the priorities for toxicological investigations. They are: (1) exposure of large industrial populations, (2) community exposures, (3) suspicious agents, (4) prolonged use of the agent, (5) planned introduction of a new agent or widened distribution of an existing agent. The most important steps to be included when an orderly system of testing is being prepared are summarized here:

1. The magnitude of usage of the agent, including its yearly production figures, utilization practice, its persistence in the environment, its exposure level and mode in the general population. An agent's ubiquity and toxicity potential should be stressed. Compounds that may be produced in low amounts but are carcinogenic or highly toxic must be given a high priority in the proposed scheme of testing.

2. An important guideline for priority selection is the chemical classifica-tion of the various substances. Arcos[33] proposed four categories of criteria for suspecting chemical compounds of carcinogenic activity:

Table III. A Partial Listing of Prioritized Organic Compounds
Considered Hazardous to the Environment and Human Health[31]

Rank	Environment	Rank	Human Health
1	Dichlorodifluoromethane	1	Benzene (chemical uses)
2	Trichlorofluoromethane	2	Ethylenimine
3	Polyhalogenated Biphenyls (Aroclor 1254)	3	Ethylene dibromide
4	Carbon Tetrachloride	4	Benzidine
5	Chloroform	5	Carbon tetrachloride
6	Hexachlorobenzene	6	Tricresyl phosphate
7	(Ethylenedinitrilo) Tetraacetic Acid Tetrasodium Salt	7	Chloroform
8	Hexachlorobutadiene	8	Allyl chloride
9	Benzene (chemical uses)	9	Polyhalogenated biphenyls (Aroclor 1254)
10	1, 3-Dichloropropane	10	Benzo(a)pyrene 1,2-dichloropropane mixture

a. *structural criteria*, based on analogies with different known types of chemical carcinogens;

b. *operational criteria*, defined by Arcos as the set of biofunctional capabilities in organic compounds likely to bring about or lead to ultimately inheritable alterations in cellular synthetic templates, or as phenomena indicating that these alterations have already occurred;

c. *"guilt by association" criterion*, which states that even though a compound may have been found noncarcinogenic under some standard conditions of testing, if it belongs to a chemical class several members of which are known to display potent and multitarget carcinogenic activity, it should be reassayed under much more stringent conditions than the standard conditions of bioassay testing;

d. *"after the fact"* criterion, in which prior human epidemiological indications are the basis for selecting chemical agents for carcinogenicity testing.

From this excellent review it is obvious that an orderly and sophisticated review of organic chemicals class by class is of great help to the toxicologist in planning a program for evaluating potential environmental toxins.

3. A recognized comprehensive toxicological testing program employs initially a battery of bioassay screening techniques (*in vitro-in vivo* bioassays) to provide data which, along with appropriate analytical procedures and monitoring of the effluents and emissions from pollution sources, are then utilized for the *in vivo* toxicological evaluation in animals, usually in rodents.[10,13,18,34,35]

The use of *in vitro* and combinations of *in vitro-in vivo* systems in qualitative toxicological testing has increased greatly in recent years. Scientists have agreed that a formula must be found for preselection of chemicals and particularly potential carcinogens before expensive, long-term *in vivo* tests are conducted. Most recently, TSCA has spelled out some of the testing procedures to be followed for newly marketed chemicals. In the battery of available *in vitro* tests, mutagenicity testing dominates the selection because it serves as an adjunct or preliminary procedure for identifying carcinogens and also, because it enables identification of mutagenic agents that could lead to chemically induced genetic diseases in an exposed population. More recently, the *Salmonella* system used for evaluation of mutagenic activity has been supplemented by various other tests to provide a better evaluation.[36,37] The strategy for screening chemicals for mutagenic activity has been recently developed to include combined *in vitro* and *in vivo* tests for sequential testing in the form of a three-tier system. Other types of *in vitro* tests include observation of changes in mammalian (human) tissue cultures and various human cell lines that are obtainable by noninvasive technique, *e.g.*, blood cells, semen, sputum and pulmonary macrophages. The *in vitro* systems are becoming more sophisticated and their value for indicating which chemicals to select for further testing has increased immensely.

4. Generally, for the *in vivo* animal investigation, three types of studies are usually considered: the *acute-high level effect* observations, such as LD_{50} or LC_{50}, the *subacute studies*, which are used for a detailed biochemical, physiological and pathological toxicity evaluation at intermediate dosage and duration; and *chronic studies*, which we feel should be planned only on the basis of all previous information and only for the most ubiquitous toxic pollutants, and should be conducted in carefully selected mammalian species at levels as close as possible to those found in the environment or work place. Weil[29] reviewed the general guidelines for experiments to predict the "Degree of Safety of a Material for Man." His outline represents the most complete statement of the testing approach that is generally followed by toxicologists everywhere. Weil's five main points are:

1. Wherever possible, use species whose biological handling of the material most closely approximates that of man.

2. Wherever practical, use several dose levels on the principle that all types of toxicological and pharmacological actions in man and animals are dose-related.

3. Effects produced at higher dose levels are useful for delineating the mechanism of action, but for any material and any of its adverse effects some dose level exists for man or animal below which this adverse effect will not appear. This biologically insignificant level can and should be set based on competent scientific judgment and using a proper safety factor.

4. Statistical tests for significance are valid only on the experimental units (*e.g.*, either litters or individuals) that have been mathematically randomized among the dosed and concurrent control groups.

5. Effects obtained by one route of administration to test animals are not *a priori* applicable to effects by another route of administration to man. The routes chosen for administration to the test animals should therefore be the same as those to be used in man. For example, food additives for man should be tested by admixture of the material in the diet of animals.

A definitive *in vivo* toxicity testing program has been further outlined by Dominguez[38] in *Guidelines of Toxic Substances Control Act.* The monograph presents detailed guidelines for the duration of acute and subactue toxicity tests in different experimental animal species, so that the meaning of the terms, including the duration of a chronic (long-term) study, can be better and uniformly understood. According to TSCA guideline, acute tests include: (1) acute oral median lethal dose (LD_{50}) in rats with single oral administration and an observation period of 14 days employing a single-dose administration of the test material; (2) acute dermal median lethal dose (LD_{50}) in rabbits (the period of application is 24 hours and the observation period is 14 days); (3) single-level acute inhalation and acute inhalation median lethal concentration (LC_{50}) in rats (exposure period is 1-4 hours). For substances on which information is not available, the acute testing program is usually performed as a normal part of the initial screening.

"Subacute Toxicity Testing" is performed to establish the nature of the toxic effects, including metabolic behavior of the substance, its bioconcentration and its retention time in the body. It should pinpoint the maximum "no-effect" dose and the Minimum Toxic Dose (MTD), as well as other pharmacokinetic parameters. The result should also provide a background for the proper design of chronic studies. The duration of the subacute studies varies from days to months. The general approach recommended for these intermediate tests is a period of about 10% of the test animal life span, *i.e.,* 90 days for a rat living 30 months, which may be compared to 7 years for a human with a 70-year life expectancy.[1]

WHY WE NEED CHRONIC (LONG-TERM) STUDIES

When a reputable toxicologist, committee or large task force evaluates the evidence provided by screening methods, the results of acute and subacute studies, and sometimes even sketchy epidemiological evidence on the potential impact of an agent on the biosphere, the question always arises whether the data are complete and satisfactory to advise authorities or the concerned industry on the safety of the product in question. Knowing that approval may result in a wide distribution of the product in the environment with its health impact on human population, the approach must be one suggested by common sense. It does not have to include a lengthy, complicated or sophisticated method of risk assessment. The paramount question should be whether the obtained data are sufficient to judge how serious an impact the material may have on public health. If serious gaps in knowledge remain, then further testing is necessary. In such a case there are only two in-depth approaches remaining to clarify the question: a chronic long-term investigation using low-level exposures in lower mammalian species selected on the basis of subclinical studies; and human epidemiological studies which may provide satisfactory correlation results but seldom a conclusive dose-effect relationship data. The best definitive guidelines should be used in selecting compounds for such expensive and long-term projects, otherwise the total resources for such a project may be wasted. A number of leaders in this area of science have dealt with various aspects of this issue. Questions that must be asked before selection of a substance for long-term studies is made should concern whether the chemical will be produced in large amounts and whether the preliminary studies indicated a high degree of toxicity. For example, will the compound become a ubiquitous contaminant affecting the public health of the nation (200 million plus people) or will it be employed on a limited scale, perhaps affecting some small segment of a working community, which could be protected if appropriate measures were employed? Is the compound ubiquitous and nondegradable in the environment, *i.e.*, does it have a long half-life in nature? Finally, from a standpoint of potential health effects, does it produce irreversible harmful changes at low levels, including, but not exclusively, carcinogenicity and teratogenicity? Such pertinent questions have been reviewed recently by Gehring *et al.*[39] and Watanabe *et al.*,[40] who suggested that metabolic activity of compounds at a higher (acute and subacute) level may be different than at a lower level normally reserved for chronic exposure effect studies. Metabolites that are produced over a longer period of time may produce harmful, irreversible changes, including induction of cancer, while other metabolites produced at high levels may not have such properties. It is clear that the shorter-term testing program cannot provide conclusive data about which chemical will cause a long-term irreversible injury in the organism, be it cancer or other irreparable functional and pathological impairment.

BASIC TECHNICAL FACTORS IN CHRONIC TOXICITY STUDY DESIGN

The technological uncertainties in using animals to assess the risk in man are numerous and must be carefully evaluated before and during the experimental design stage when planning chronic studies. Some of the more important factors are described in the following paragraphs.

Species

The choice of animal species is often the most difficult question in developing the experimental design of any study. Several factors must be considered, such as similarity in metabolic pattern of a substance between experimental species and man, similar toxicological manifestations including tumorigenesis, longevity, body weight and period of gestation. Primates are chosen more often than any other mammalian species to predict more closely the manifestations of toxic material in man. However, a selection of primates as the experimental animal is at least 100-fold more costly than when rodents are used. In addition, there are instances in which primate data do not parallel behavior of certain agents in man. The best example is the experience with Isoniazid (INH).[41] In monkeys, Isoniazid was very well tolerated at doses of 100 μg/kg. In dogs, on the other hand, doses of 20 μg/kg produced clonic convulsions, respiratory failure and death. When a sufficient dose of pyridoxine was given, the symptoms in dogs were completely reversed. For comparison, the LD_{50} in rats was found to be 1,670 mg/kg. Because of the information it was suggested that a large dose of INH could be given safely to humans. In some cases this may be so; nevertheless, many of the people treated with higher doses had undesirable side effects, such as a peripheral neuritis. It was eventually realized that the reason for the differences of the effects in monkeys was that they acetylate INH almost completely, whereas dogs cannot. Some humans are rapid inactivators and acetylate the compound, whereas many others could not acetylate it completely. Another example is an antirheumatic drug (Phenylbutazone), which is metabolized slowly in man but in mice, rats, rabbits, dogs, guinea pigs and horses it is rapidly eliminated from the body.[42] Of particular interest in the experimental animal selection process are pathways that are present in most species but missing in others. For example, dogs cannot acetylate primary amines and cats cannot convert phenols to glucuronides.[42] A necessary step that must not be overlooked by a toxicologist is to determine whether some human data are already available and then base the selection of animal species on a similar metabolic and physiologic response.

As suggested earlier, subacute metabolic data are of utmost importance for an optimal approach toward design of a chronic study. A good example of this is presented in Table IV, which summarizes the duration of action and metabolism of hexobarbital; it also makes the case for dog as the experimental animal. On the other hand, an example of a case of a complete miss of experimental species choice is griseofulvin, which was given to dogs over a two-year period and reported to be completely nontoxic. It was realized later than griseofulvin does not absorb from the gastrointestinal tract in the dog.[41] Another example is the case of mercury: rats or mice do not have much value for studies of neurotoxicity of methyl mercury

Table IV. Species Differences in Duration of Action
and in Metabolism of Hexobarbital[42]

Species[a]	Duration of action (min)	Biologic Half-Life (min)	Plasma Level of Hexobarbital on Awakening (μg/ml)	Relative Enzyme Activity (μg/gm/hr)
Mouse	12 ± 8	19 ± 7	89 ± 31	598 ± 184
Rabbit	49 ± 12	60 ± 11	57 ± 12	196 ± 28
Rat	90 ± 15	140 ± 54	64 ± 8	134 ± 51
Dog	315 ± 105	260 ± 20	19 ± 4	36 ± 30
Man		360	20	

[a]The dose of barbiturate was 100 mg/kg for mouse, rabbit and rat and 50 mg/kg for dog.

because they rapidly metabolize this compound to a less toxic inorganic form. In this case, nonhuman primates or cats should have been used because they metabolize methyl mercury similarly to man.[43] In most carcinogenic assays the choice species are rats, mice or hamsters, because a large number of test animals are needed to observe dose effects and because the animals must be observed for most of their life span.

To summarize, a great deal has been written about selection of the most appropriate species for a chronic study design as was pointed out by some of the examples. It is true that many more rats can be used than dogs or monkeys. However, even in case of rodents the resulting statistics are inadequate to extrapolate to a population of more than 200 million people. Use of large mammals such as primates, dogs, cats, miniature pigs and miniature goats is well advised, particularly when there is a need for repeated sampling over a significant period of time. Hematological, excretionary, functional and many other parameters are usually required in the experimental design of a long-term study, even though the final results in large mammals often are to be interpreted more on a qualitative than quantitative basis. Another advantage of larger mammals is that base line observations during the pretreatment period can be easily obtained; as a consequence, the animals can be used as their own controls. Such data improve the quality and precision of biological effects observations. Various authors feel that extrapolating to human biological effects from animal data is complicated by the use of rodents as experimental animals because, according to Sterling,[44] "rodents are much further removed phylogenetically from the human animal than are dogs or monkeys." According to Wood,[45] monkeys are unquestionably closer phylogenetically to humans than dogs and rodents, yet common ancestors are not known for the three orders—Primates, Carnivora and Rodentia—and their interrelationship is not generally agreed on other than that they all are placental mammals. From the fossil records it appears that primates and rodents are more closely related than either is to carnivora. The recent review of the origin of earlier rodents suggested that rodents may have originated from primates. Since their primate ancestors lived some 70×10^6 years ago, present rodents and primates would not be closely related; however, it is estimated that the common ancestor of primates and carnivores lived even earlier. If this is correct the tests on rodents should give as good indications of human reaction as would tests on dogs, but this may be just a conjecture because some other factor, such as the life span of the species, may be the more important consideration.

Life Span

The life spans of humans and experimental animals are not in proportion and this must be considered in the design of long-term effect studies. Krasovskii[46] estimated that in the case of carcinogenic doses, which can cause effects decades after exposure, the time period for an experimental model must correspond to at least 5-6 times the life span of the model. He further postulated that the average life span of the rat (approximately 2.5 years) corresponds only to 15-17 years of human life. It is necessary then to use the most appropriate toxicological techniques to study the impaired sites and functions of an organism, yet the series of tests available are often nonspecific. This is particularly true when studying degenerative diseases of aging, such as chronic nephritis, arteriosclerosis or emphysema. There is an inherent difficulty in reproducing such effects in experiment animals with a short life span. Many studies demonstrate that the dependability of results is questionable at best and that reliable data could not be obtained in shorter-lived species such as rodents because of the latency period needed for development of pathological lesions. Dogs and other higher mammals have been used successfully in a number of investigations, particularly in the fields of radiation, anticancer drug research, effects of pesticides and, more recently, effects of environmental pollution. In listing some examples, Dogherty et al.[47] have found that the important biological end points observed in dogs exposed to two isotopes of radium (^{226}Ra and ^{228}Ra) were quite comparable to those observed in humans and that deposition patterns of plutonium (^{239}Pu) in dogs more closely approximated the distribution in humans than did the deposition patterns in rats. Goldman and Della Rosa[48] have reported on the dynamics of ^{90}strontium metabolism in dogs exposed from intrauterine time to adulthood. Their results indicate that the long-term dynamics of ^{90}Sr metabolism are similar in some respects to that of ^{90}Sr in man (Table V).

Table V. Comparison of Strontium and Calcium Metabolism
in Standard Man (70 kg) and "Standard" Beagle (10 kg)[48]

Parameter	Strontium Beagle	Strontium Man	Calcium Beagle	Calcium Man	Beagle/Man per kg Sr	Beagle/Man per kg Ca	Sr/Ca Beagle/Man
Intake, mg/day	5.0	1.6	3,500	1,200	21.9	20.4	1.1
Absorption, mg/day	0.3	0.3	400	450	7.0	6.2	1.1
Endogenous Fecal Excretion, mg/day	0.2	0.04	370	150	35.0	17.2	2.0
Urinary Excretion, mg/day	0.10	0.24	30	250	2.9	0.84	3.5
Sweat, mg/day		0.02		50			
Hair, mg/day	0.001		0.04				
Total in Plasma, mg	0.02	0.08	40	350	1.8	0.8	2.2
Exchangeable Pool, mg	0.4	1.2	500	5,000	2.3	0.7	3.3
Accretion Rate in Bone, mg/day	0.04	0.12	90	500	2.3	1.26	1.8
Content in Bone, mg for Sr, g for Ca	64.0	250.0	140	1,040	1.8	0.94	1.9
Strontium/Calcium Ratio In bone mg [90] Sr/g Ca	0.6	0.24					2.5
Observed Ratio (bone/diet)	0.4	0.18					2.2

Leach et al.[49] subjected dogs, monkeys and rats to the inhalation of natural uranium oxide (UO_2) dust at a concentration of 5 mg/m^3, 6 hr/day, 5 days/wk for up to 5 years. The lung and tracheobronchial lymph node data for dogs and monkeys in this study suggested that there may be a radiation hazard in these tissues at or below the recommended TLV. These dogs were exposed for periods of up to 5 years with little evidence of injury; however, 2-6 years postexposure a high percentage of the dogs developed pulmonary neoplasia.[50] Hobbs et al.,[51] in comparing the toxicity of inhaled cerium oxide ($^{144}CeO_2$) in dogs, mice and hamsters found much lower doses to be effective in reducing the life span of the dogs as compared to the rodents on a μCi/kg basis. The rationale suggested for this observation is that rodents consistently have a much shorter lung retention of inhaled, relatively insoluble particles than do dogs. In another study, Hobbs et al.[52] have been unable to demonstrate pulmonary neoplasms in hamsters exposed to lung burdens of $^{239}PuO_2$ equivalent to those shown to produce neoplasms in dogs three years postexposure. It is suggested that the latent period for tumor development may not be proportional to life span, in which case hamsters may not live long enough to develop pulmonary neoplasms.

Dogs have also proved useful in the evaluation of risk resulting from insult imposed by a variety of pesticides. One study that illustrates the potential value of long-term dog studies of pesticide toxicity for assessment of human health effects is that of Deichmann et al.[53] In this study dogs were dosed for 14 months with DDT. The pesticide concentration in the blood and fat during this time was comparable to levels in healthy occupationally exposed workers.

Another large mammal that has been used extensively in long-term toxicity testing is the miniature swine. This animal has proved to be especially useful in the study of ^{90}Sr-induced leukemia. Clarke et al.,[54,55] Howard and Clarke[56] and Ragan et al.[57] have published a series of reports describing various aspects of a study in which miniature swine were dosed orally with ^{90}Sr at levels from 1-3,100 μCi/day for 7-10 years. These reports pointed out that bone marrow damage rather than osteosarcomas may have been the limiting factor in chronic oral toxicity of ^{90}Sr and that the latent period for leukemia induction was considerably shorter than that for development of osteosarcoma. Figure 3 shows that the cumulative mortality in weaned offspring at the lower dose levels became significantly different from controls only after 10 years of exposure.[54] Another animal model that has not been used extensively, but which has proved to be quite useful in a number of long-term studies, is the cat. A group of papers reporting on various metabolic and effect parameters observed after oral administration of ^{89}Sr in cats have been published by Berman and Stara,[58] Nelson, et al.,[59,60] and Stara and Wolfangel.[61] Topics discussed in these papers include: (1) uniquely sensitive tissues to radiation-induced lymphoproliferative or myeloproliferative disorders during growth and development of the cat; (2) assessment of radiation dose rate delivered to the bone marrow and other organs in cats of different ages; and (3) the effective leukomogenic dose of ^{89}Sr in cats vs the effective ^{90}Sr dose in dogs and miniature swine.

Another area in which dogs have proved to be useful test subjects concerns the quantification of effects resulting from cigarette smoking. Auerbach et al.[62] have reported elevated heart weight:body weight ratios in smoking dogs as well as pulmonary fibrosis and emphysema, which were similar to conditions reported in humans. In another paper, Auerbach et al.[63] have demonstrated a thickening of myocardium arteriole walls in smoking dogs. Thickening increased with the duration of smoking and with the number of cigarettes smoked and also in dogs smoking nonfilter as opposed to filter cigarettes. Hammond et al.[64] found that the responses observed in dogs following smoking closely parallel those observed in man in that the types of histologic changes produced in the lung parenchyma were the same in both species. There was a dose-response relationship and the degree of damage to the lung parenchyma increased with duration of cigarette smoking in both species.

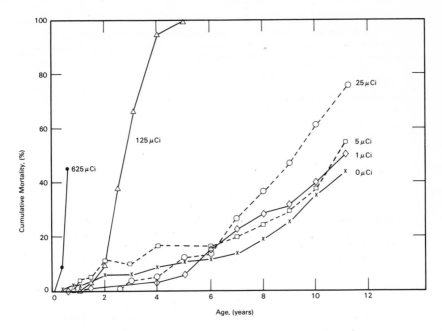

Figure 3. Cumulative mortality in weaned offspring of dams fed $^{90}Sr^{54}$.

A monograph reporting a major, 9-year study of effects of air pollutants in beagles was recently prepared and edited by Stara *et al.*[65] Since auto exhaust represents a complex mixture of interacting chemical substances of varying toxicity, which may pose a potential hazard to human health, the objective of the study was to examine the harmful effects of long-term exposures to environmentally realistic levels of whole exhaust emissions as well as of their individual components. The major goal of the study was to assess the irreversible chronic effects of the tested air pollutants in dogs to provide an indication of potential effects in similarly exposed human populations. Tables VI and VII show the functional impairment of the respiratory system that occurred in the dogs exposed to sulfur oxides, nitrogen oxides, raw auto exhaust and simulated smog following a period of 68 months of exposure to peak urban air pollution levels. The data show the difference in total lung dead space between controls and all experimental groups exposed to raw exhaust, sulfur oxides, mixture of auto exhaust and sulfur oxides, nitrogen dioxide and mixture of irradiated exhaust and sulfur oxides. Similarly, the tables show the changes in the mean total lung capacity at the end of a 68-month exposure and 2 years later at sacrifice.

Morphologic examination of dog lungs in this study by light microscopy, scanning electron microscopy (SEM) and transmission electron microscopy

Table VI. Total Dead Space of Control Group Compared
with Exposed Groups Two Years after Termination of Exposure

	Total Dead Space (ml)
Control	45.1
R	72.0
SO_X	69.6
$R + SO_X$	66.6
NO_2 –Low	65.3
$I + SO_X$	62.4

Table VII. Mean Total Lung Capacity (ml) of Each Group of Beagles at End of
Exposure (TE, column 1), Two Years after End of Exposure (2 yr, column 2),
Change in Volume Between TE and 2 yr (column 3), Predicted Value from
External Controls (Column 4) and the Difference Between Predicted
and 2 yr (Column 5)

Groups	1 TE	2 2-yr	3 Change	4 Predicted	5 Difference
Control	979	1,095	116	1,028	67
R	1,188	1,374	186	1,156	218
I	1,087	1,235	148	1,168	67
SO_X	1,144	1,450	306	1,168	282
$R + SO_X$	1,139	1,248	109	1,096	152
$I + SO_X$	1,166	1,272	106	1,191	81
NO_2 –High	1,183	1,342	159	1,108	234
NO_2 –Low	1,175	1,262	87	1,132	130

(TEM) revealed that exposure-related lesions included centriacinar emphysema,
characterized by enlargement of distal respiratory bronchioles and proximal
alveolar ducts, and hyperplasia of nonciliated bronchiolar cells. The centriacinar emphysema lesions were most severe in the dogs exposed to oxides of
nitrogen or to oxides of sulfur and were often associated with fenestrations
or prominent increases in the number and size of interalveolar pores. By contrast, hyperplasia of nonciliated bronchiolar cells was most severe in dogs
exposed to raw exhaust alone or with oxides of sulfur. The summary of some
of these effects is presented in Table VIII. Foci of ciliary loss with and without squamous metaplasia were occasionally observed in the trachea and bronchi. The emphysematous and hyperplastic lesions correlated with functional
impairment reported for these dogs. The major toxicological implications

Table VIII. Summary of Results of Pulmonary Lesions (Evaluated by SEM)

Lesions	Overall Differences Between Groups[a]	Groups Significantly Different from Control[b,c]	Groups Significantly Different from Control in Dorsal Region[d]	Groups Significantly Different from Control in Ventral Region[d]
Ciliary Loss Without Squamous Metaplasia	NS[e]	SO_x, NO_2 high	None	None
Ciliary Loss with Squamous Metaplasia	NS[e]	R + SO_x	None	None
Nonciliated Bronchiolar Cell Hyperplasia	p < 0.005	R + SO_x, R, I, NO_2 high, SO_x, I + SO_x	R, R + SO_x	R + SO_x, R, I, NO_2 high, NO high
Interalveolar Pores	p < 0.01	NO high, SO_x, R + SO_x, I + SO_x, R	NO high, SO_x, R + SO_x, I + SO_x, R	NO high, SO_x, R + SO_x, I + SO, R, NO_2 high

[a]Kruskal-Wallis test.
[b]Groups significantly different from the control group with more severe lesions (P ≤ 0.05, Mann-Whitney U Test).
[c]Groups listed from most to least severe lesions.
[d]Groups significantly different from the control group with more severe lesions in the dorsal or ventral regions (p ≤ 0.05, Mann-Whitney U Test).
[e]No significant differences found (p > 0.10).

were: (1) the production of permanent lung damage at much lower concentrations of pollutants than previously reported, and (2) the apparent lack of additive or synergistic effects between oxidant gases and sulfur oxides. The irreversible nature of bronchiolar cell proliferations was also demonstrated.

Krasovskii[46] reported that the average life spans for 70 species of mammals were linearly related to body weight (Figure 4). All species are plotted on the graph in life span segments from A - I as follows: (A) shrew, mouse (domestic), mouse (field), flying mouse; (B) chipmunk, golden hamster; (C) squirrel, rat, hedgehog, guinea pig; (D) skunk, agouti, jackal, porcupine, rabbit, cat, rhesus monkey, raccoon, polar fox, mongoose, marmot, common hare, lynx; (E) coyote, badger, civet, cat, tapir, nutria, dog guepard, wolf, reindeer, goat, jaguar, seal, baboon; (F) antelope, dolphin, sheep, leopard, chimpanzee, hyenia, fallow deer, man, llama, alpaca, puma, orangutang; (G) pig, lion, wild boar, grizzly bear, donkey, deer, tiger, gorilla, zebra; (H) moose, camel, buffalo (Asiatic), buffalo (African), polar bear, horse, bull, bison, giraffe, hippopotamus; (I) rhinoceros, elephant. This observation is in agreement with other experiment reports. The dimensions of the blood vessels and the body length:width ratio were in a linear correlation with mammalian body weight.

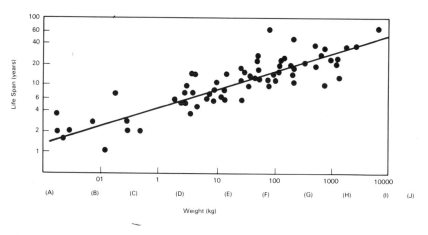

Figure 4. Variation of the average life span of some mammals with increased body weight.

An excellent example of life span being a critical factor in chronic effects is cadmium. The "Itai-Itai" disease diagnosed in older Japanese women represents actually the first discovered human disease due to an environmental agent. There has been a tremendous amount of research done on the effects of cadmium in experimental animals and yet none of the experiments had predicted that this condition would occur only in older women after menopause, whose diet was low in calcium and who had delivered four or more children. Studies in rats dealing with end points such as carcinogenicity and hypertension have not predicted that the kidney is the target organ in humans because of cadmium's accumulative properties (half-life is 20-30 years). Not only is the Cd body distribution in man opposite to that found in shorter animal experiments (about 70% in liver and 20% in kidney), but also toxicologic manifestations are different, particularly after long-term/low-level exposure as in the case of Itai-Itai disease.[66] Selecting the optimal time for conducting chronic toxicological experiments is all important and there is no strict scientific basis for predicting optimal time to observe effects of chemicals given daily to animals. It is necessary to compare the effective doses of substances computed on the basis of the dose-effect-time relationship for various periods of intoxication, which may range from one week to the life span of the animals.

Influence of Age

Krasovskii,[46] Kostial,[67] Kello[68] and others have reported significant variations of effects based on age sensitivity. Metabolic and toxicity studies of several metals have shown a very high intestinal absorption in sucklings due to milk diet, higher whole body retention, higher blood levels and higher accumulation in various organs such as brain occuring in sucklings in comparison to adult animals (Table IX). There were specific differences reported in the pharmacokinetics of metals[69] indicating that the early neonatal age is a critical period for metal accumulation and toxicity in an organism.

Table IX. Influence of Age and Milk Diet on Lead, Cadmium, Mercury and Manganese Absorption in Rats

	Absorption, % of oral dose[a]					
			Older Animals			
Radio-isotope	Age Weeks (n)	Sucklings	Age Weeks (n)	Milk Diet	Age Weeks (n)	Standard Diet
^{203}Pb	1(18)	52.05±1.65	6(24)	22.9 ±2.15	6(20)	0.40±0.09
^{115m}Cd	1(24)	25.61±2.0	6(10)	6.86±0.24	6(10)	0.49±0.02
^{203}Hg	1(23)	38.21±1.50	18(11)	6.73±0.83	18(11)	0.99±0.05
^{54}Mn	1(18)	39.9 ±1.5	6(15)	6.4 ±0.4	6(11)	0.05±0.004

[a]Percentage oral dose in whole body 6 days after administration. Values are presented as arithmetic mean ± SE; number of animals in parentheses.

Krasovskii[46] summarized data on animal sensitivity to 119 substances and proposed that the ratio of differences of young and adult organisms was 2.4±0.23, *i.e.*, the young animals are about two and a half times more sensitive than adults. It can be concluded that age may be a significant variable to be considered when designing studies.

Body Weight

Krasovskii[46] also discussed the importance of body weight. He proposed a formulation whereby the logarithms of mammalian biological parameters are in a linear correlation with the logarithms of body weight. He expressed this rule by a parabolic function, $x = ay^B$, or by the straight line equation:

$$\log x = \log z + b \log y$$

The linear correlation between the indices of toxicity and the body weight of various animal species was a particular case of this body weight rule. In analyzing further the relationship between species differences and the action of several hundred chemical compounds, it was shown that the regularities of the comparative sensitivity of the animals to 80-85% of the substances can be equally characterized by a straight line equation.

Sex Differences

A number of reports on effects of chemicals and drugs indicate that there is often a difference between males and females and how differently they handle various materials. Table X, taken from Coulston,[41] shows the comparative toxicity of parathion in rats, proving that female rats are much more sensitive to parathion when compared to males. However, to add to the confusion of species selection, this sex difference has not been observed in either monkeys or man.

Table X. Oral Toxicity (LD_{50}) of Parathion in Rats

Male	LD_{50}	13.0 mg/kg
Female	LD_{50}	3.6 mg/kg

Another example is Norbormide, which when given to the white albino rats resulted in an LD_{50} dose for females being 5.3 mg/kg vs the male rats, which tolerated more and had an LD_{50} dose of 15 mg/kg.[41] Brodie[42] reported that when male and female rats were given hexobarbital as an anesthetic agent, the duration of narcosis for female rats was about four times as long as for the males. A comparison of the plasma levels indicated that the drug was meta-bolized much more rapidly in males than in females. However, this sex difference in the reaction to hexobarbital was not observed in some of the other animal species, including mice, guinea pigs, rabbits and dogs. Kello *et al.*[70] reported on the influence of sex on whole body retention of dietary calcium using orally administered radioactive cadmium chloride tracer. Male rats retained less cadmium than females (Table XI).

Table XI. Gastrointestinal Absorption of Cadmium-115m[a]

	Male	Female
Control	0.23	0.45
Diet (1.1%)	0.05	0.05
Low Calcium	0.30	0.60
Diet (0.3%)	0.05	0.06
High Calcium	0.08	0.12
Diet (2.4%)	0.01	0.01

[a]Percentage of whole body retention 7 days after per os administration.

These data clearly point out that sex might influence the level of cadmium deposition in the body. Krasovskii[46] summarized the sexual differences of white rats with respect to 149 compounds. Most of the animals did not exceed 2-3 times the limit and only in case of certain phosphoroorganic compounds, where the females were three to four times more sensitive, was there a wider spread.

Strain

Different strains of experimental animals have been shown to produce biological effects that were different in time or incidence. Such observations were made particularly in inbred strains of mice or rats. One such example reported by Brodie[42] is antipyridine, which is oxidized by inbred strains of rats at much different rates. In the M-520 rat strain, the half-life of this drug was 114 minutes as compared to 290 minutes for the Buffalo rat strain. Therefore, the metabolism rate of this drug may be based on heredity rather than other environmental factors. Coulston[41] also reported that the duration of action of hexobarbital differed among several inbred strains of mice; however, extremely uniform responses were observed in individual mice within a certain strain. At the same time, members of a strain that had not been inbred varied considerably in their reaction to the drug. Another example is the lethal effect of Norbormide (McNeil Laboratories), which is used widely to kill brown rats but not mice, grey rats or roof rats. Table XII demonstrates the acute toxicity of Norbormide to the genus Rattus and shows how strain and sex affect the biological behavior of this compound.

Table XII. Acute Toxicity of Norbormide in Genus Rattus[41]

Species[a]	Variety	Route	Sex	No. Rats	LD$_{50}$ (mg/kg)	95% Conf. limits
Norvegicus	Domestic albino rat	i.v.	Female	50	0.63	0.56 - 0.71
			Male	40	0.65	0.60 - 0.71
		i.p.	Female	50	3.5	2.8 - 4.4
			Male	40	4.4	3.5 - 5.5
		p.o.	Female	30	5.3	4.4 - 6.5
			Male	40	15.0	12.0 - 19.0
	Wild Norway rat	p.o.	Female	21	10.0	7.1 - 15.0
			Male	24	13.0	11.0 - 15.0
			Mixed	45	11.5	10.0 - 13.0
Rattus	Wild roof rat	p.o.	Mixed	30	52.0	47.0 - 57.0
Hawaiiansis	Wild Hawaiian rat	p.o.	Mixed	15	10[a]	

[a]Three of six rats died at 10 mg/kg; three of three died at 30, 100 and 300 mg/kg.

Dose

The importance of dose was mentioned earlier in this chapter. A very careful selection of dose must be made in any experimental design, particularly for chronic studies. Much of the needed information is obtained from the subchronic testing, since several dose levels are usually used for both acute and subacute toxicity studies. Also, metabolic data, including the rate of absorption and excretion, are usually made available. One of the dose levels chosen should be the dose that produces frank signs of toxicity but does not cause mortality or major functional changes. Often this so-called maximal tolerable dose (MTD) is determined as a part of a subchronic range-finding study. Under present standard protocol the MTD is usually the highest dose used for carcinogenic evaluation of chemicals. The lowest selected dose should not produce *any* evidence of toxicity; under the experimental conditions this dose is the so-called "no effect" dose. In selecting additional dose levels consideration should be given to the actual human exposure levels present in the particular environment and to the margin of safety desired for the particular compound. Whenever possible, the dose level selected should be the actual measured environmental concentration of the compound in air, water and other environmental media. Dixon[71] stated that the greatest problem in extrapolation from animals to humans is the appropriate selection of dose and that the best clinical data are obtained when the experimental MTD expressed in mg/kg is used for the more sensitive of the large animal species, be it dogs or monkeys. Clinical introduction of a new anticancer agent at a dose of 1/10 of the MTD in the most sensitive higher mammalian species carries a statistical risk of about 3%; *i.e.,* about 3 of every 100 new drugs used at initial dosage in a clinical study may be expected to produce some toxic effect in man.

Diet

Diet is another essential factor in design of chronic studies. It must be free of all toxic impurities and must meet all nutritional requirements. Dietary intake of inorganic substances and organic nutrients might influence the metabolism and toxicity of metals. Milk, which was recommended in the past as an antidote agent in lead poisoning because of its high calcium content, was actually found to greatly enhance lead absorption.[72] However, although diet in a long-term study can be carefully selected and prepared, it may still differ in structure and content from the human diet and the resulting differences in absorption rates may have to be dealt with. Kostial and Kello[73] have studied the lead bioavailability in rats fed "human diets." In that instance, the lead absorption values in rats were similar to values observed in humans. It would be desirable to perform similar experiments using a variety of chemical compounds and animal species. Such data are extremely important because the design of human experiments, for example in the case of balance studies, has to take into consideration the nutrition habits of various population groups, especially young children. Krasovskii[46] reported a sixfold difference in the individual sensitivity of children with respect to the action of fluorine and nitrates in drinking water. This presents a convincing case that the "safe" levels of environmental chemical pollution should be determined by animal experiments to be harmless not only for an "average" human but also for all subgroups of the population, including children, the aged and chronically ill.

CONCLUSIONS

There are many other pieces of the scientific puzzle that can be enumerated. In this review we listed those that, according to our experience, appear most crucial. They must be carefully considered when preparing an experimental design for a chronic toxicological study program, which may cost a great deal in terms of money and personnel.

The toxic substances law provides the framework for dealing with environmental pollutants on a scale never before possible. The U.S. Environmental Protection Agency, which is responsible for carrying out this task, is dedicated to this goal and has public support. The question is whether practical application of the requirements and goals of this act can fulfill theoretical expectations. Many government officials and scientists across the U.S. do not yet have a clear view of the act's practical application, yet at present, the practical aspects are the most important issues that need to be clarified. As to the controversies among the scientists, decisions must be reached and followed for this important plan to receive proper direction. To protect future generations we must develop a practical approach. Figure 5 clearly shows a paradox between the contribution of health sciences in this century to public health, population growth and longevity on the one hand, and the hazards that this and future generations may face from an uncontrolled chemical pollution of the environment and the work place on the other.

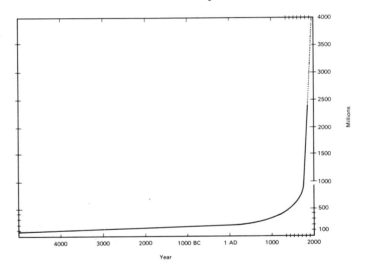

Figure 5. World population growth (5000 B.C. to 1950 A.D.).[74]

Obviously, the greatest potential danger is overpopulation and lack of food and energy resources to sustain it. The main reason, in our opinion, to protect the environment and control the hazardous toxicants even at low levels, is because the earth is a closed system, not dissimilar from a test tube or agar plate, in which bacterial population growth flourishes, reaches its peak, and eventually begins to die off. We must strive to provide the best balance and environmental quality for optimal growth and development of the population. For this reason, studies of health risk assessments due to chemical contaminants of the environment and their subsequent control measures are of the utmost importance. Science, government, industry and the public sector must urgently consider this most serious issue and expedite its solution. Our time is running out.

REFERENCES

1. Friedman, L. "Symposium on the Evaluation of the Safety of Food Additives and Chemical Residues: II. The Role of the Laboratory Animal Study of Intermediate Duration for Evaluation of Safety," *Toxicol. Appl. Pharmacol.* 16:498 (1970).
2. Truhaut, R. "Can Permissible Levels of Carcinogenic Compounds in the Environment be Envisaged?", *Ecotoxicol. Environ. Safety* 1:31 (1977).
3. Toxic Substances Control Act (October 11, 1976).
4. Muul, I., A. F. Hegyeli, J. C. Dacre and G. Woodard. "Toxicological Testing Dilemma," *Science* 193:834 (1976).
5. Falk, H. L. "The Toxic Substances Control Act and *In Vitro Toxicity Testing*," *In Vitro* 13:676 (1977).
6. Culliton, B. J. "Toxic Substances Legislation: How Well are Laws Being Implemented?", *Science* 201:1198 (1978).
7. Maugh II, T. H. "Chemical Carcinogens: how Dangerous are Low Doses?" *Science* 202:37 (1978).
8. Muir, W. R. "Prevention of Occupational Cancer - Toward an Integrated Program of Governmental Action," (Discussion of part IV.) *Ann. N.Y. Acad. Sci.* 271:491 (1976).
9. "Toxic Substances," Council of Environmental Quality, Washington, DC (1971).
10. Zapp, J. A., Jr. "Extrapolation of Animal Studies to the Human Situation," *J. Toxicol. Environ. Health* 2:1425 (1977).
11. Gehring, P. J., and G. E. Blau. "Mechanisms of Carcinogenesis: Dose Response," *J. Environ. Pathol. Toxicol.* 1:163 (1977).
12. *Drinking Water and Health* (Washington, DC: National Academy of Sciences, 1977).
13. Magee, P. N. "Extrapolation of Cellular and Molecular Level Studies to the Human Situation," *J. Toxicol. Environ. Health* 2:1415 (1977).
14. Goth, R., and M. F. Rajewsky. "Persistance of 06-Ethylguanine in Rat Brain DNA: Correlation with Nervous System Specific Carcinogenesis by Ethylnitrosourea," *Proc. Nat. Acad. Sci. U.S.A.* 71:639 (1974).
15. Margison, G. P., and P. Kleihues. "Chemical Carcinogenesis in the Nervous System. Preferential Accumulation of 0^6-Methylguanine in Rat Brain Deoxyribonucleic Acid During Repetitive Administration of N-Mythyl-N-Nitrosourea," *Biochem. J.* 148:521 (1975).
16. Nicoll, J. W., P. F. Swann and A. E. Pegg. "Effect of Dimethylnitrosamine on Persistence of Methylated Guanines in Rat Liver and Kidney DNA," *Nautre* 254:261 (1975).
17. Jones, H. B., and A. Grendon. "Environmental Factors in the Origin of Cancer and Estimation of the Possible Hazard to Man," *Food Cosmet. Toxicol.* 13:251 (1975).
18. Commoner, B. "Chemical Carcinogens in the Environment," in *Identification & Analysis of Organic Pollutants in Water*, L. H. Keith, Ed. (Ann Arbor, MI: Ann Arbor Science Publishers, Inc., 1976), p. 49.

19. Who Expert Committee. "Prevention of Cancer," *Tech. Rep. Series* 276:1 (1964).

20. Saffiotti, U. "Scientific Bases of Environmental Carcinogenesis and Cancer Prevention: Developing an Interdisciplinary Science and Facing its Ethical Implications," *J. Toxicol. Environ. Health* 2:1435 (1977).

21. Rall, D. P. "Difficulties in Extrapolating the Results of Toxicity Studies in Laboratory Animals to Man," *Environ. Res.* 2:360 (1969).

22. Hueper, W. C. "Public Health Hazards from Environmental Chemical Carcinogens, Mutagens and Teratogens," *Health Phys.* 21:689 (1971).

23. Thiede, T., and B. C. Christensen. "Bladder Tumours Induced by Chlornaphazine," *Acta Med. Scand.* 185:133 (1969).

24. Herbst, A. L., H. Ulfelder and D. C. Poskanzer. "Adenocarcinoma of the Vagina. Association of Maternal Stilbestrol Therapy with Tumour Appearance in Young Women," *N. Engl. J. Med.* 284:878 (1971).

25. Johnson, F. L., J. R. Feagler, K. G. Lerner, P. W. Majerus, M. Siegel, J. R. Hartmann and E. D. Thomas. "Association of Androgenic-Anabolic Steriod Therapy with Development of Hepatocellular Carcinoma," *Lancet* 2:1273 (1972).

26. Baum, J. K., F. Holtz, J. J. Bookstein and E. W. Klein. "Possible Association between Benign Hepatomas and Oral Contraceptives," *Lancet* 2:926 (1973).

27. Creech, J. L., and M. N. Johnson. "Angiosarcoma of Liver in the Manufacture of Polyvinylchloride," *J. Occup. Med.* 16:150 (1974).

28. Nelson, N. "The Chlorethers-Occupational Carcinogens: A Summary of Laboratory and Epidemiology Studies," *Ann. N. Y. Acad. Sci.* 271: 81 (1976).

29. Weil, C. S. "Guidelines for Experiments to Predict the Degree of Safety of a Material for Man," *Toxicol. Appl. Pharmacol.* 21:194 (1972).

30. Selikoff, I. J. "Perspectives in the Investigation of Health Hazards in the Chemical Industry," *Ecotoxicol. Environ. Safety* 1:387 (1977).

31. Stephenson, M. E. "An Approach to the Identification of Organic Compounds Hazardous to the Environment and Human Health," *Ecotoxicol. Environ. Safety* 1:39 (1977).

32. Lepkowski, W. "Extrapolation of Carcinogenesis Data," *Environ. Health Perspect.* 22:173 (1978).

33. Arcos, J. C. "Criteria for Selecting Chemical Compounds for Carcinogenicity Testing: an Essay," *J. Environ. Pathol. Toxicol.* 1:433 (1978).

34. Ames, B. N. "The Detection of Chemical Mutagens with Enteric Bacteria," in: Chemical Mutagens. Principles and Methods for their Detection, Vol. 1, A. Hollaender, Ed. (New York: Plenum Publishing Corp. 1971), p. 267.

35. Ames, B. N. "A Bacterial System for Detecting Mutagens and Carcinogens," in: *Mutagenic Effects of Environmental Contaminants*, H. E. Sutton and M. I. Harris, Eds. (New York: Academic Press, 1972), p. 57.

36. Ames, B. N., J. McCann and E. Yamasaki. "Methods for Detecting Carcinogens and Mutagens with the Salmonella/Mammalian Microsome Mutagenicity Test," *Mut. Res.* 31:347 (1975).

37. Legator, M. S., and S. Zimmering. "Integration of Mammalian, Microbial and Drosophila Procedures for Evaluating Chemical Mutagens," *Mut. Res.* 29:181 (1975).

38. Dominguez, G. *Guidebook: Toxic Substances Control Act* (Cleveland, OH: CRC Press, 1977).

39. Gering, P. J., P. G. Watanabe and C. N. Park. "Resolution of Dose-Response Toxicity Data for Chemicals Requiring Metabolic Activation: Example - Vinyl Chloride," *Toxicol. Appl. Pharmacol.* 44:581 (1978).

40. Watanabe, P. G., J. A. Zempel and P. J. Gehring. "Comparison of the Fate of Vinyl Chloride Following Single and Repeated Exposure in Rats," *Toxicol. Appl. Pharmacol.* 44:391 (1978).

41. Coulston, F. "Qualitative and Quantitative Relationships Between Toxicity of Drugs in Man, Lower Mammals, and Nonhuman Primates," *Proc. Conf. on Nonhuman Primate Toxicol.*, C.O. Miller, Ed. (1966), p. 31.

42. Brodie, B. B. "Part VI. Difficulties in Extrapolating Data on Metabolism of Drugs from Animal to Man," *Clin. Pharmacol. Therapeut.* 3:374 (1962).

43. Task Group on Metal Toxicity. "Consensus Report," in *Effects and Dose-Response Relationships of Toxic Metals*, G. F. Nordberg, Ed. (Amsterdam: Elsevier Scientific Publishing Co., 1976), p. 58.

44. Sterling, T. D. "Difficulty of Evaluating the Toxicity and Teratogenicity of 2,4,5-T from Existing Animal Experiments," *Science* 174:1358 (1971).

45. Wood, A. E. "Interrelations of Humans, Dogs, and Rodents," *Science* 176:437 (1972).

46. Krasovskii, G. N. "Extrapolation of Experimental Data from Animals to Man," *Environ. Health Perspect* 13:51 (1976).

47. Dougherty, T. F., B. J. Stover, J. H. Dougherty, W. S. S. Jee, C. W. Mays, C. E. Rehfeld, W. R. Christensen and H. C. Goldthorpe. "Studies of the Biological Effects of ^{226}Ra, ^{239}Pu, ^{228}Ra(MsTh), ^{228}Th(RdTh) and ^{90}Sr in Adult Beagles," *Rad. Res.* 17:625 (1962).

48. Goldman, M., and R. J. Della Rosa. "Studies on the Dynamics of Strongium Metabolism under Conditions of Continuous Ingestion to Maturity," in *Strongtium Metabolism, J. M. Lenihan et al., Eds.* (New York: Academic Press, 1967), p. 181.

49. Leach, L. J., E. A. Maynard, H. G. Hodge, J. K. Scott, C. L. Yulie, G. E. Sylvester and H. B. Wilson. "A Five-Year Inhalation Study with Natural Uranium Dioxide Dust (UO_2). I. Retention and Biologic Effect in the Monkey, Dog and Rat," *Health Phys.* 18:599 (1970).

50. Leach, L. J., C. L. Yulie, H. C. Hodge, G. E. Sylvester and H. B. Wilson. "A Five-Year Inhalation Study with Natural Uranium Oxide Dust. II. Post-Exposure Retention and Biological Effects in the Dog and Rat," *Health Phy.* 25:239 (1973).

51. Hobbs, C. H., R. O. McCellen and S. A. Benjamin. "Toxicity of Inhaled ^{144}CeO$_2$ in Immature, Young Adult and Aged Syrian Hamsters," II. *Inhalation Toxicol. Res. Instit. Ann Rep.* 1974-1974, Lovelace Fn., Albuquerque, NM (1974a).

52. Hobbs, C. H., J. A. Mewhinney, D. A. Slauson, R. O. McClellen and J. J. Meglio. "Toxicity of Inhaled ^{239}PuO$_2$ in Immature, Young Adult and Aged Syrian Hamster," II. *Inhalation Toxicol. Res. Inst. Ann Rep.* 1973-1974, Loelace Fn., Albuquerque, NM (1974b).

53. Deichmann, W. B., W. E. MacDonald, A. G. Beasley and D. Cubit. "Subnormal Reproduction in Beagle Dogs Induced by DDT and Aldrin," *Ind. Med. Surg.* 40:10 (1971).

54. Clarke, W. J., E. B. Howard and P. L. Hackett. "Strontium-90 Induced Neoplasia in Swine," in *Delayed Effects of Bone-Seeking Radionuclides*, C. W. Mays *et al.* Eds. (Salt Lake City: University of Utah Press, 1969), p. 263.

55. Clarke, W. J., R. F. Palmer, E. B. Howard and P. L. Hackett. "Strontium-90: Effects of Chronic Ingestion on Farrowing Performance of Miniature Swine," *Science* 169:598 (1970).
56. Howard, E. B., and W. J. Clarke. "Induction of Hematopoietic Neoplasms in Miniature Swine by Chronie Feeding of ^{90}Sr," *J. Nat. Cancer Inst.* 44:21 (1970).
57. Ragan, H. A., P. L. Hackett, B. J. McClanhan and W. J. Clarke. "Pathologic Effects of Chronic ^{90}Sr Ingestion in Miniature Swine," in *Research Animals in Medicine*, L. T. Harmison, Ed., DHEW Pub. No. (NIH) 72-333 GPO Washington, DC (1973), p. 919.
58. Berman, E., and J. F. Stara. "Qualitative Evaluation of the Embryonic Skeleton in Cats," in *Radiation Bioeffects Summary Report* U.S. HEW (January-December 1968), p. 73.
59. Nelson, N. W., R. G. Wolfangel and J. F. Stara. "Placental Transport of Strontium and Calcium in the Felis Domestic," in *Radiation Bioeffects Summary Report*, U.S. HEW (January-December, 1968), p. 63.
60. Nelson, N. S., J. F. Stara and R. C. Wolfange."^{85}Sr Distribution and Weights of Organ Systems in Chronically-Labelled Cats," in *Radiation Bioeffects Summary Report*, U.S. HEW (January-December, 1968), p. 70.
61. Stara, J. F., and R. G. Wolfangel. "Metabolism of ^{85}Sr in Cats," in *Radiation Bioeffects Summary Report*, U.S. HEW (January-December, 1968), p. 66.
62. Auerbach, O., E. C. Hammond, D. Kirman and L. Garfinkel, . "Emphysema Produced in Dogs by Cigarette Smoking," *J. Am. Med. Assoc.* 199:241 (1967).
63. Auerbach, O., C. E. Hammond and L. Garfinkel. "Thickness of Walls of Myocardial Arterioles in Relation to Smoking and Age," *Arch. Environ. Health.* 22:20 (1971).
64. Hammond, E. C., D. Kirman and L. Garfinkel. "Effects of Cigarette Smoking on Dogs," *Arch. Environ. Health* 21:740 (1970).
65. Stara, J. F., D. L. Dungworth, W. S. Tyler and J. G. Orthoefer. "Study of Chronic Effects of Air Pollutants in Beagles," U.S. Environmental Protection Agency, in press.
66. Friberg, L., M. Piscator, G. F. Nordberg and T. Kjellstrom. *Cadmium in the Environment*, 2nd Ed. (Cleveland, OH: CRC Press, 1974).
67. Kostial, K., I. Simonovic and M. Pisonic. "Lead Absorption from the Intestine in Newborn Rats," *Nature (London)*, 233:564 (1971).
68. Kello, D., and K. Kostial. "Influence of Age and Milk Diet on Cadmium Absorption from the Gut," *Toxicol. Appl. Pharmacol.* 40:277 (1977).
69. Kostial, K., D. Kello, S. Jugo, I. Rabar and T. Maljkovic. "Influence of Age on Metal Metabolism and Toxicity," *Environ. Health Perspect.* 25:81 (1978).
70. Kello, D., D. Dekanic and K. Kostial. "Influence of Sex and Dietary Calcium on Intestinal Cadmium Absorption in Rats," *Arch. Environ. Health*, in press.
71. Dixon, R. L. "Problems in Extrapolating Toxicity Data for Laboratory Animals to Man," *Environ. Health Perspect.* 13:43 (1976).
72. Kello, D., And K. Kostial. "The Effect of Milk Diet on Lead Metabolism in Rats," *Environ. Res.* 6:355 (1973).
73. Kostial, K., and D. Kello. "Bioavailability of Lead in Rats Fed "Human Diet," *Bull. Environ. Contam. Toxicol.*, in press.
74. Coriell, L. L. "The Scientific Responsibilities at Issue," *In Vitro* 13:632 (1977).

CHAPTER 5

PULMONARY ALVEOLAR MACROPHAGE FUNCTION: SOME EFFECTS OF CIGARETTE SMOKE*

J. Bernard L. Gee, Bruce R. Boynton,
Atul S. Khandwala and G.J. Walker Smith

Departments of Medicine and Pathology
Yale University School of Medicine
New Haven, Connecticut 06510

ALVEOLAR MACROPHAGE FUNCTION

The most obvious function of the alveolar macrophage (AM) is as a phagocytic scavenger. This capability does serve the critical lung defense function against microbial attack and clears or degrades materials whose aerodynamic and physicochemical characteristics are such that they may be deposited in either the small airways or terminal bronchoalveolar units. Since cigarette smoke contains both "particles and vapors" that reach these sites in the lung, the effects on AM function are of interest. This account will summarize certain aspects of macrophage smoke interactions, namely macrophage mobilization in human cigarette smokers and the effects of cigarette smoke *in vitro* on rabbit AM. Finally, brief reference will be made to another less obvious but important AM function—its secretory function. Reviews of AM function have appeared recently[1-4] and details may be found therein.

---

*Supported by NIH Grant HL 19237

MACROPHAGE MOBILIZATION BY CIGARETTE SMOKE

Following dust inhalation, large numbers of AM appear in the distal parts of the bronchoalveolar tree of experimental animals. Similarly, in man, bronchoscopic pulmonary lavage yields about three times more cells from the lungs of cigarette smokers than are obtained from nonsmokers.[5,6] We have seen recently the open lung biopsies of two heavy smokers in whom no other disease entity has been apparent. The light and electron microscope slides are reproduced in Figures 1 and 2. In Figure 1, the large mass of macrophages lying within the lumen of a small airway are readily apparent. In Figure 2, the AM is shown to contain large numbers of residual bodies, which represent the relics of phagocytosed particulate matter. The striking high prevalence of AM within these airways and of such residual bodies provides clear evidence of the heightened scavenger activity in the lung of smoking man. This histologic evidence of increased lung air space macrophage content indicates that the large numbers of AM obtained by bronchoalveolar lavage of smokers' lungs result from an increased AM content as opposed to decreased AM adherence to lung tissue. Additionally, this striking increase in AM content will be important in later considering the heightened secretory function of AM noted in cigarette smokers.

METABOLIC EFFECTS OF CIGARETTE SMOKE ON ALVEOLAR MACROPHAGES

Normal Metabolic Pathways

Detailed accounts of AM metabolism can be found elsewhere.[2,3] For our purposes, three general features will serve as the basis for the account of our work in this area. First, phagocytosis of bacteria by rabbit AM is associated with increased oxygen (O_2) consumption and with increased conversion of labeled (^{14}C) glucose to $^{14}CO_2$, both by the pentose shunt (^{14}C-1-glucose) and the combination of glycolysis and mitochondrial pathways (^{14}C-6-glucose). Some components of these pathways are shown in Figure 3, where hydrogen peroxide (H_2O_2) generated from molecular O_2 is linked to the pentose shunt by a glutathione shuttle system. It should be noted that in addition to reduced glutathione (GSH), both glutathione reductase and peroxidase are SH-containing enzymes. In the latter enzyme, selenium as opposed to sulfur is present at the active site. This glutathione system serves as antioxidant disposal role both for H_2O_2 and lipid hydroperoxides.[7]

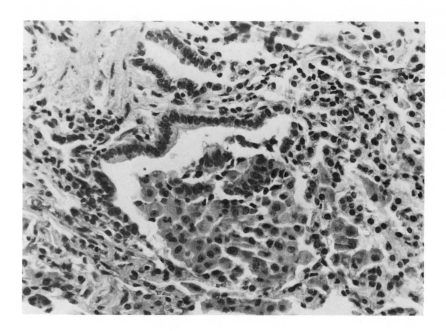

Figure 1. A small airway lined by bronchiolar epithelium is filled by an abnormal number of macrophages. Hematoxylin and eosin (X250).

Second, the reduction of molecular O_2 is not as simple as depicted in Figure 3. In addition to H_2O_2, superoxide anion ($O_2^-\cdot$) and singlet excited oxygen, $^1O_2^*$, are formed by AM. The rates of formation of these three unstable intermediates of O_2 reduction are probably increased during phagocytosis.[2,8,9] Additionally, superoxide anion and H_2O_2 interact by the Haber-Weiss reaction to form the unstable hydroxyl free radical (OH·):

$$O_2^-\cdot + H_2O_2 \rightarrow OH\cdot + OH^- + {}^1O_2^*$$

The latter radical is capable of injuring cell membrances and can cause autolysis in the polymorphonuclear leukocyte. The concentrations of H_2O_2 and O_2^- are affected by the glutathione system and by the enzymes catalase and superoxide dismutase (SOD). Thus, these enzymes also serve as antioxidants.

Third, the AM are relatively rich in mitochondria. AM function at the relatively high O_2 tension of the lung and under normal circumstances critically require oxidative phosphorylation as an energy source. Energy sources in macrophages, namely glycolysis and oxidative phosphorylation, however, are capable of adaptation to variations in O_2 tension. These may be seen by comparing the enzyme activities of the glycolytic pathway

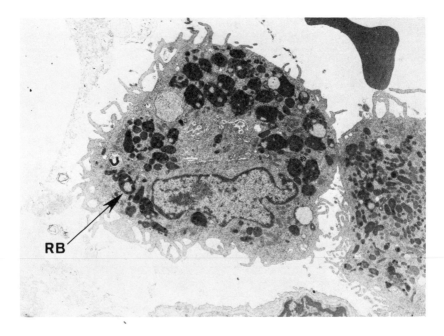

Figure 2. Electron micrograph of one AM and a portion of cytoplasm of an adjacent macrophage. Note the many intracytoplasmic residual bodies (RB). Alveolar wall is present at the bottom of photo (X5860).

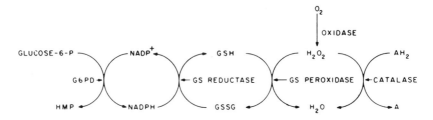

Figure 3. Peroxidative metabolic pathways. GSH = reduced glutathione; GSSG = oxidized glutathione; G6PD = glucose 6-phosphate dehydrogenase; HMP = hexose monophosphate shunt. AH_2 represents unknown hydrogen donors for catalase. GS reductase and GS peroxidase refer to glutathione enzymes.

and of the mitochondria in the relatively anaerobic peritoneal macrophage and the largely aerobic pulmonary alveolar macrophage. Additionally, variations in O_2 tension induce changes in these same groups of enzymes in alveolar macrophages grown in tissue culture.

EFFECTS OF CIGARETTE SMOKE *IN VITRO*

Energy Metabolism

We undertook some studies of the effects of aqueous extract of filtered cigarette smoke on both phagocytosis and the related metabolic pathways. We suspected that while the oxidant effects are important, such a complex mixture as cigarette smoke would have many more points of action on AM metabolism than had hitherto been defined. A detailed account of this work is published elsewhere.[11] An aqueous extract of cigarette smoke (SE) in Krebs-Ringer phosphate pH 7.4 was prepared as described by Powell and Green.[4] To determine whether the effects of SE were due to oxidants, we followed the approach of Green's group in which the antioxidants, reduced glutathione (GSH) or cysteine, are added to the experimental system immediately prior to the addition of SE. The findings were as follows:

1. Particle uptake was measured using heat-killed *S. auraus*. SE (0.1 μl) produced a 50% inhibition of uptake. All our subsequent studies were designed to elucidate the mechanism(s) impairing particle uptake. Most importantly, this impaired uptake was not prevented by either antioxidant. Thus, while there are important effects of oxidizing substances in SE, they are not the major determinant of impaired particle uptake.

2. Oxidant effects of SE on three SH-containing enzymes are indicated in Table I. The effect on the glutathione enzymes provides a reasonable explanation of the observation that SE also impaired the production of $^{14}CO_2$ from ^{14}C-1-glucose in phagcytosing AM, in which the pentose shunt is stimulated by a glutathione-dependent H_2O_2 disposal pathway.[2,12] This view is supported by the observation that both antioxidants restored the SE impairment of $^{14}CO_2$ production from ^{14}C-1-glucose to control levels.

3. The nonoxidant effects of SE appear of greater importance as a source of impaired particle uptake.

Table I. Oxidant Effects of SE on Enzyme Activities[a]

	+SE	+SE +GSH[b]	+SE +Cy[b]
Gluthathione Peroxidase	75	100	100
Glutathione Reductase	48	100	100
Glyceraldehyde-3-Phosphate Dehydrogenase	82	100	100

[a]Data expressed as % control.
[b]SE = smoke extract; GSH = reduced glutathione; Cy = cysteine.

We next showed that, as judged by the rates of conversion of ^{14}C-6-glucose to $^{14}CO_2$, glycolysis is impaired by SE, an effect not prevented by the antioxidants. This effect on glycolysis could not be ascribed to an effect on key enzymes involved in the Embden-Myerhof pathway. Hexokinase and phosphofructokinase were unaffected by SE. The small oxidant effect of SE on glyceraldehyde-3-phosphate dehydrogenase seems too small to modify glycolysis.

Evidence that the impairment of $^{14}CO_2$ production from ^{14}C-6-glucose derives from impaired mitochondrial function may now be summarized:

1. SE doubled the rate of lactate formation by resting AM.
2. $^{14}CO_2$ production from ^{14}C-1-pyruvate, ^{14}C-1-4-succinate and ^{14}C-1-acetate were all sharply dimished by SE (Figure 4). These effects could depend on impaired substrate entry into the AM, separate impairment of the appropriate enzymes or, more likely, on other impairments of mitochondrial function.
3. SE evoked a fall of intracellular ATP concentrations in the AM to 40% of the control levels. For comparison, we found KCN at 10^{-4} M concentration diminished ATP levels to 61% of control. In the AM, which is rich in mitochondria, ATP generation derives largely from oxidative phosphorylation.

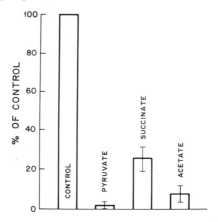

Figure 4. Effect of Se on $^{14}CO_2$ production from ^{14}C-labeled acetate, pyruvate and succinate. AM (1×10^7) were incubated in a total volume of 2 ml of GRPS containing 0.25 ml of homologous serum at $37°C$ for one hour. The experimental flasks had 0.75 ml SE. The corresponding control flasks had 0.75 ml GRPS. At the end of the incubation 1 μCi of ^{14}C-1-acetate, or ^{14}C-1-pyruvate or ^{14}C-1-4-succinate were added to all the flasks and the production of $^{14}CO_2$ during the next hour was measured as described in the methods section. The results presented as a percentage of control (cells in absence of any additive = 100%) represent the average ± standard error of four experiments done in triplicate.

Taken as a whole, the simplest and most economical explanation of these three sets of observations is that SE impairs mitochondrial function in some undefined manner. The mechanism(s) involved were not fully investigated. We did show, however, that SE inhibited the activities of succinic acid dehydrogenase and cytochrome oxidase by 50% and 16%, respectively. The inhibition of cytochrome oxidase could result from CN^- present in SE. We did not, however, seek spectroscopic evidence for this mechanism, since it seemed more likely that there was more than one process involved in the effects. Our data certainly implicate mitochondrial malfunction and specifically demonstrate that the energy metabolism of AM is sharply diminished by SE.

Other *In Vitro* Effects of Cigarette Smoke

Green and co-workers[4] have devoted considerable effort to studies of AM behavior in the presence of cigarette smoke and have emphasized the role of oxidant materials. Additionally, employing somewhat different conditions, they report that acrolein is an important toxic agent. They also define effects on glyceraldehyde-3-phosphate dehydrogenase and on protein synthesis. Additional effects of cigarette smoke on AM *in vitro* are summarized elsewhere.[1,4]

EFFECTS OF CIGARETTE SMOKE *IN VIVO* IN MAN

There are obvious problems associated with short-term *in vitro* studies of the effects of smoke on AM. The preparation of the smoke, its inherent instability, the nature of the media containing the AM and the short course of such experiments all impose limitations in the interpretation and variability in the results of such experiments. Additionally, even *in vivo* studies of smoking in experimental animals are open to question. Thus, studies of the behavior of AM harvested from man by bronchoscopic bronchoalveolar lavage are of much more direct interest. Some of the observations on such human AM have been itemized in recent reviews.[1-3] There are several features of AM from cigarette smokers that suggest the cells are activated in a manner similar to that observed following immunization by procedures which operate by delayed hypersensitivity mechanisms. These features include:

1. increased glass adherence and spreading;
2. increased QO_2 and glucose oxidation; and
3. increased secretion of lysozyme and increased content of certain enzymes such as angiotensin convertase.[6]

Additionally, Senior and colleagues[13] have recently reported that the neutral proteinase, elastase, is actively secreted in tissue culture by AM derived in cigarette smoking man. This particular enzyme is potentially of great importance because it can attack elastin in a native, insoluble extra-cellular form. This structural protein provides the basis for lung elastic recoil which, in turn, both maintains the patency of the small peripheral airways and contributes to the expiratory driving pressures.

These features, suggesting an increased activity of AM from cigarette smokers, contrast with the *in vitro* studies described above. They empha-size the adaptive capacity of this cell[2] but indicate that such adaptations include the potential for tissue destruction. The increased activity of the individual cell becomes particularly important when one recalls the im-pressive numbers of AM found in the lungs of cigarette smokers.

ACKNOWLEDGMENTS

This work was supported by NIH Grant HL 19237.

REFERENCES

1. Mason, R.J. "Metabolism of Alveolar Macrophages," *Respiratory Defense Mechanisms*, J.D. Brain, D.F. Proctor and L.M. Reid, Eds. (New York: Marcel Dekker, Inc., 1977), p. 893.
2. Gee, J.B.L., and A.S. Khandwala. "Motility, Transport and Endocytosis in Lung Defense Cells," in *Respiratory Defense Mechanisms*, J.D. Brain, D.F. Proctor and L.M. Reid, Eds., (New York: Marcel Dekker, Inc., 1977), p. 927.
3. Allison, A.C. "Mechanisms of Macrophage Damage in Relation to the Pathogenesis of some Lung Diseases," in *Respiratory Defense Mechanisms*, J.D. Brain, D.F. Proctor and L.M. Reid, Eds. (New York: Marcel Dekker, Inc., 1977), p. 1075.
4. Green, G.M., G.J. Jakab, R.B. Low and G.S. Davis. "Defense Mechan-isms of the Respiratory Membrane," *Amer. Rev. Resp. Dis.* 115: 479 (1977).
5. Harris, J.O., G.N. Olsen, J.R. Castle and A.S. Maloney. "Comparison of Proteolytic Enzyme Activity in Pulmonary Alveolar Macrophages and Blood Leukocytes in Smokers and Nonsmokers," *Am. Rev. Resp. Dis.* 111:579 (1975).
6. Hinman, L.M., C.A. Stevens, R.A. Matthay, H.Y. Reynolds and J.B.L. Gee. "Lysozyme and Angiotensin Convertase Activity in Human Alve-olar Macrophages: the Effects of Cigarette Smoking and Pulmonary Sarcoidosis," *Am. Rev. Resp. Dis.* 117 (part 2): 67 (1978).
7. Khandwala, A. and, J.B.L. Gee. "Linoleic Acid Hydroperoxide: Impaired Bacterial Uptake by Alveolar Macrophages, A Mechanism of Oxidant Lung Injury. *Science* 182: 1364 (1973).
8. Gee, J.B.L., C.L. Vassallo, P. Bell, J. Kaskin, R.E. Basford, and J.B. Field. Catalase-Dependent Peroxidative Metabolism in the Alveolar Macro-phage During Phagocytosis, *J. Clin. Invest.* 49: 1280 (1970).

9. Gee, J.B.L., and A.S. Khandwala. "Oxygen Metabolism in the Alveolar Macrophage: Friend or Foe? *J. Reticuloendothel. Soc.* 19: 229 (1976).

10. Salin, M.L. and J.M. McCord. "Free Radicals and Inflammation. Protection of Phagocytosing Leukocytes by Superoxide Dismutase," *J. Clin. Invest.* 56: 1319 (1975).

11. Gee, J.B.L., A.S. Khandwala and B.R. Boynton. "Aqueous Extract of Cigarette Smoke. Effects on Metabolism and Phagocytosis by Rabbit Alveolar Macrophages," in *Movement, Metabolism and Bactericidal Mechanisms of Phagocytes*, R. Rossi, P.L. Patriarca and D. Romeo, Eds. (Padua, Italy: Piccin Editore, 1977), p. 385.

12. Gee, J.B.L., J. Kaskin, M.P. Duncombe and C.L. Vassallo. "The Effects of Ethanol on Some Metabolic Features of Phagocytosis in the Alveolar Macrophage," *J. Reticuloendothel. Soc.* 15: 61 (1974).

13. Rodriguez, R.J., R.R. White, R.M. Senior and E.A. Levine. "Elastase Release from Human Alveolar Macrophages: Comparison Between Smokers and Nonsmokers," *Science* 198: 313 (1977).

CHAPTER 6

ALTERATION IN HOST-BACTERIA
INTERACTION BY
ENVIRONMENTAL CHEMICALS

Donald E. Gardner

Environmental Protection Agency
Health Effects Research Laboratory
Research Triangle Park, North Carolina 27711

INTRODUCTION

Environmental toxicology seeks to elucidate cause and effect relationships between test substances and the exposed individual. Biological responses can often be evoked by increasing the concentration of the test substance and by examining for physiological, biochemical or morphological effects. However, at lower levels of exposure, numerous subtle trends may be observed which are not statistically significant and, therefore, are often disregarded. These subtle effects may or may not be significant to the host. Therefore, it becomes important to develop exquisitely sensitive model systems to aid the investigator in understanding the effects taking place at the target site. This can be difficult, however, since the establishment of a disease state is very complex. There is the host, the causative agent and several often unidentified modifiers, such as stress, preexisting disease, dietary deficiency or smoking, which can modify and complicate the etiology and pathogenesis of the disease.[1-5]

Traditionally, environmental toxicologists have been concerned primarily with studying the host and the causative agent, being careful to avoid any additional factors that might modify the individual's response. Their primary goal was to identify the agent that might cause a toxic response and to provide guidelines for appropriate control. Recently, however, investigators have also begun to incorporate into toxicological studies various biological, chemical and physical factors, which can modify the response of the host to the test substance.[6-9] Such interaction studies are more relevant since their design more closely reflects the natural human condition of multiple exposure to a number of components, which might include combinations of pollutant, microbes and other environmental stresses, *e.g.,* heat, cold and exercise. In addition, these models permit the toxicologist to perform dose response studies at lower levels, which mimic those found in the ambient environment.

For many years, it has been apparent that infectious diseases are not the result of a simple one-to-one cause and effect relationship, but instead are a reflection of several different influences acting simultaneously on the host. For the establishment of an infectious disease, not only is there a requirement for a specific microorganism, but additional factors are necessary to render a particular individual vulnerable to microbial invasion.

This becomes very evident when one considers that the lungs have a remarkable capacity to cope with inhaled microorganisms, as indicated by the virtual sterility of the lower respiratory tract, even though it is in the presence of a constant microbial assault from the external environment.[10,11] Particles deposited in the tracheobronchial region are, within hours, transported out of the lung by the mucociliary system.[12] The primary defense of the deep lung, however, is the alveolar macrophage, which removes the particles more slowly, but is capable of inactivating and killing viable microorganisms within two to four hours.[13-16] Any alteration in these self-cleaning actions gives the invading microorganism an enhanced opportunity to avoid host defenses and to multiply uninterruptedly. Recently, there have been several excellent reviews on defense mechanisms of the respiratory tract and how infectious diseases and pollutants can cause defects in function ranging from impairment of particle transport to the dysfunction of the biochemical events of phagocytosis.[2,4,10,12,14,17,18]

CRITERIA FOR MODELS FOR HOST DEFENSES

The suppression of pulmonary defense mechanisms by environmental chemicals would be expected to result in prolonged bacterial viability in the lung, thus increasing the risk of disease. These factors provide the basis for studying inhalation toxicology with unique animal model system, which places air pollutants in perspective as one of the environmental modifiers in the course of an experimental infection. This model system would reflect the summation of various deleterious changes within the lung, which may include edema, reduced phagocytic or bactericidal activity of the pulmonary macrophages, altered ciliary activity, inflammation and immunosuppression. Such a model should be valid across species, allow for reproducible data, be a sensitive indicator of the effects of a wide range of pollutants, be supported by mechanistic experimental information and be capable of epidemiological investigations.

ENHANCEMENT OF LABORATORY INDUCED INFECTION

The infectivity model reflects the complex interaction of the host, pollutants and microbes and fulfills the above criteria. Briefly, the animals are selected randomly to be exposed to either the test substance or to clean air. After the cessation of the exposure, the animals from both groups are combined and exposed for approximately 15 minutes to an aerosol of infectious microorganisms. Immediately after the bacterial challenge, selected animals from each group are removed and sacrificed, and the number of microbes initially deposited in the lung is determined using standard microbiological techniques. The number of viable microbes deposited in the lungs immediately after the cessation of the bacterial challenge can range from 200-4,000 colony-forming units. Within this range, there is no relationship (r = 0.002) between the number of colony-forming units of the resulting mortality. This is clearly illustrated in the scattergram (Figure 1) that shows no relationship between these variables.[19] This can be repeated at various time periods to follow the clearance or growth of the microbes in the lung. The remainder of the animals are returned to clean filtered air and the mortality rate in each group is observed over a 15-day period. By exposing the animals to the test substance and subsequently challenging them with an aerosol of viable microorganisms, the model measures the propensity of toxicants to enhance the susceptibility of the host to infectious disease.

This infectivity model system has been employed successfully with a variety of animal species, including the squirrel monkey, hamster, rat and mouse.[20-25] Numerous microorganisms, including *Streptococcus pyogenes, Escherichia coli, Diplococcus pneumoniae, Klebsiella pneumoniae, Salmonella typhimurium* and Influenza virus have been employed. To be effective in the model system, the microorganisms should fulfill certain criteria: (1) they must be able to multiply and infect susceptible tissue; (2) the mortality rate must be low in animals not exposed to the toxicant; (3) small variations in dose or virulence of the microorganisms should not greatly influence the mortality rate, (4) the organism must be able to be aerosolized; and (5) a method must be available for quantifying the bacterial dose. Comparison studies have shown that using *Streptococcus pyogenes,* Group C, in the model results in a high level of reproducibility and

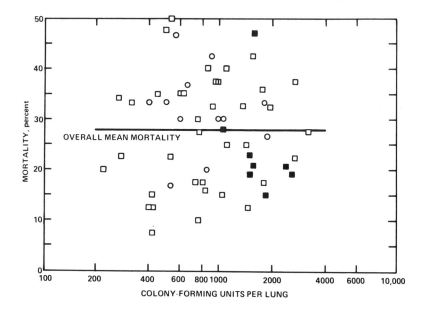

Figure 1. Relationship between percent mortality and the number of viable microorganisms deposited in the lung of mice. The correlation coefficient of 0.002 illustrates that no relationship existed between these variables. ■,□ and 0 represent the data from experiments that used 100, 40 and 30 control mice, respectively.

sensitivity. The mortality rate in control animals (exposed only to the microorganism) is usually 15-20% and reflects the natural susceptibility of the host to this infectious agent.

EFFECTS OF GASEOUS CHEMICALS

Inhalation of a number of noxious gases has been shown to increase mortality in this system. Detailed studies on the influence of the exposure mode on the toxicity of nitrogen dioxide (NO_2) have been conducted recently for several different concentrations (0.94 - 52.6 mg/m^3) [23,26,27] (Figure 2). The predicted length of exposure needed to produce a significant 20% enhancement in mortality varies from approximately 6,150 hr for 0.94 mg NO_2/m^3 (0.5 ppm) to 30 minutes for 26.3 mg NO_2/m^3 (14 ppm). Also, comparisons were made between responses observed from intermittent exposure (7 hr/day, 7 day/wk) to those animals exposed continuously.[27] When the data were evaluated on the basis of total c x t (concentration x time), the percent mortality for the two exposure modes was essentially the same (Figure 3).

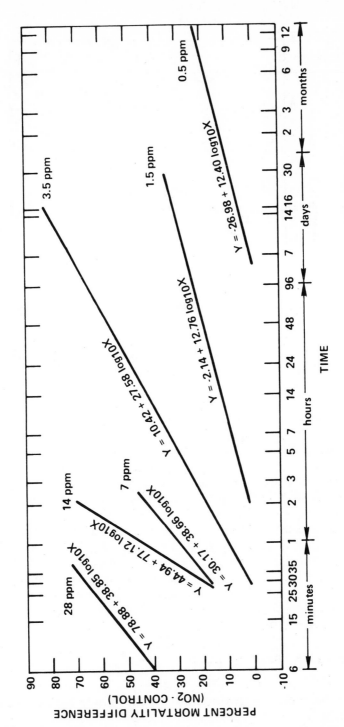

Figure 2. Percent mortality of mice versus length of continuous exposure to various concentrations of NO₂ prior to an aerosol challenge with streptococci.

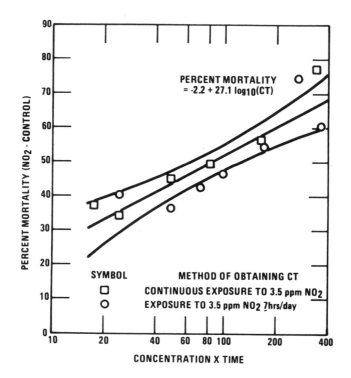

Figure 3. Percent mortality of mice from bacterial pneumonia following continuous and intermittent exposure to 3.5 ppm NO_2 versus C x T.

Ozone (O_3) exposure for three hours at concentrations greater than 0.196 $\mu g/m^3$ (0.1 ppm) also significantly enhances the mortality rate from streptococcal pneumonia.[28] To demonstrate adverse effects at 0.157 or 0.196 mg/m^3 O_3 (0.08 or 0.1 ppm), it is necessary to administer the bacterial aerosol during the pollutant exposure (Table I).[19]

A single pollutant effect can be further enhanced by adding more stressors. Exercise can increase both O_3 and NO_2 toxicity.[9,28] Combinations of O_3 with NO_2 or with H_2SO_4 produce additive effects. (Figure 4).[29,30] In the latter case, a statistically significant increase in respiratory infections occurred only when exposure to the oxidant immediately preceded that of the acid. The adverse effects of sulfuric acid (H_2SO_4) exposure are also potentiated by concurrent particulate exposure.[31]

EFFECTS OF TRACE METALS

Enhancement of laboratory-induced streptococcal pneumonia in mice was studied following separate two-hour exposure to soluble aerosols of

Table I. Infectivity Studies on Mice Exposed to 0.08 ppm Ozone [a]

Treatment Group	Number of Viable Microorganisms Per Lung Initially Deposited (cfu/lung)	Mortality (%)	Difference in Mortality (O_3 - Control) (%)
Control	879 ± 54	19.2 ± 1.8[a]	5.4[b] ± 3.0
Ozone (0.08 ppm)	926 ± 82	24.6 ±2.2	

[a] Data mean values ± SE. Control and O_3 treatment bacterial deposition means are based upon 25 and 19 mice, respectively. Mortality means are from 12 replicate experiments with 40 mice/group.

[b] Mean difference in mortality is significantly greater than zero ($p < 0.05$).

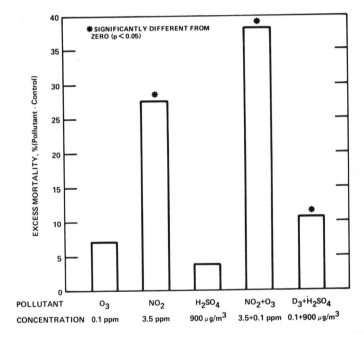

Figure 4. Enhancement of toxic effect as measured by excess mortality due to exposure to combination of pollutants.

nickel chloride (215-667 μg Ni/m^3), nickel sulfate (368-516 μg Ni/m^3), cadmium chloride (74-1,990 μg Cd/m^3), cadmium sulfate (111-164 μg Cd/m^3) and manganese chloride (900-7,560 μg Mn/m^3).[16,24,32,33] Manganous manganic oxide was tested at concentrations of 500-5,200 μg/m^3 (Figure 5). Inert particulates, such as carbon black (5.0 mg/m^3) and iron oxide (2.5 mg/m³), did not exhibit any significant effect. Utilizing the infectivity model results, the ranking of toxicity for these metals is cadmium > nickel > manganese, which agrees well with the data from *in vitro* alveolar macrophage toxicity testing using the same metals.[34] Cook *et al.*[35] have demonstrated that intravenous administration of lead acetate (2 mg/ 100 g) or cadmium acetate (0.6 mg/100 g) enhances approximately 1,000-fold the susceptibility of rats to challenge with *E. coli.* Mice treated with subclinical doses of lead nitrate for 30 days showed greater susceptibility (increased mortality) to *Salmonella typhimurium* than controls that received no lead.[36]

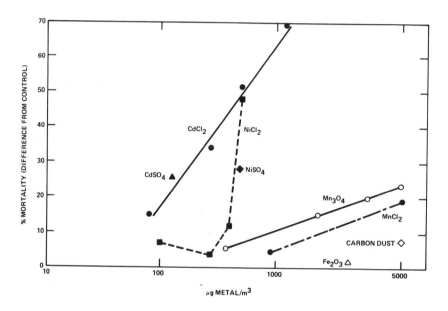

Figure 5. Percent mortality in mice resulting from inhalation of particulates of various metals and carbon dust prior to an aerosol challenge of viable streptococci.

ALTERATIONS IN CELLULAR DEFENSES

When a dose-response relationship has been determined for mortality, studies of the mechanisms of action can be better designed. Investigations into the major defense systems of the lung, *i.e.,* alveolar macrophages and mucociliary transport, have been shown to be altered by the same pollutants that enhance mortality in the infectivity model.[10,37-40] The mucociliary escalator is primarily responsible for clearing any particulates that may impact on the mucous layer. The effectiveness of the mucociliary escalator depends on normal functioning of the ciliated epithelium. Alterations of directed ciliary beating and mucous viscosity and composition may increase the host's susceptibility to infection. Numerous techniques have been used to show that industrial particulates, gaseous pollutants and cigarette smoke can cause mucociliary dysfunction.[40-43]

ALTERATIONS IN CILIARY ACTIVITY

In our laboratory, we have measured the effects of gaseous and particulate environmental chemicals on ciliary activity by combining an *in vivo* exposure with an *in vitro* assay. In this sytem, hamsters are exposed to the test substance, and at various times after the exposure, the animals are sacrificed and the bioassay for ciliary activity performed *in vitro* by use of a stroboscope.[40] This system offers the unique advantage of allowing the measurement of the ability of animals to recover from the insult.

Cd, Ni and H_2SO_4 can effectively depress and can even be lethal to ciliary activity at various concentrations and exposure lengths. Hamsters exposed for two hours to concentrations as low as 50 $\mu Cd/m^3$ of air displayed significant reduction in ciliary beating frequency. Experiments designed to determine the rate of recovery following exposure to Cd indicated that the effect was still evident even after the animals had been allowed to breathe clean air for six weeks.[40] Similarly, statistically significant reduction in ciliary activity occurred at concentrations as low as 100 $\mu g/m^3$ of Ni (Figure 6).[41] H_2SO_4 (900 $\mu g\ H_2SO_4/m^3$) also significantly reduced the tracheal ciliary activity; however, a three-hour exposure to ozone (0.1 ppm) failed to depress ciliary activity (Figure 7). Thus, the tracheal organ culture technique appears to be an excellent and sensitive model for assessing the effects of pollutants on the mucociliary escalator.

ALTERATIONS IN ALVEOLAR MACROPHAGES

Numerous toxicological studies have been performed which clearly indicate that both gaseous and particulate environmental chemicals can interfere with the functioning of the alveolar macrophage. *In vitro* tests have shown that a number of metals, such as VO_3^-, Cd^{2+}, Ni^{2+}, Mn^{2+} and Cr^{3+}, can impair deep lung clearance by being cytotoxic to these phagocytic cells (Figure 8).[34,37,46]

When alveolar macrophages from exposed animals are studied, the effect of inhalation of environmental chemicals such as O_3, NO_2, Cd^{2+} and Ni^{2+} becomes evident.[16,24,32,39,45] There can be a diminution of the *in vivo* phagocytic properties of the pulmonary macrophages resulting in a marked reduction in number of bacteria which are ingested by the

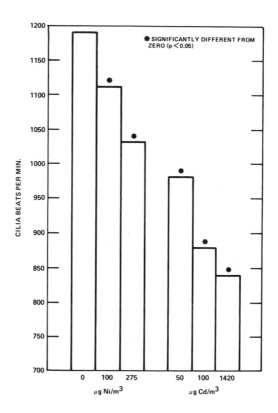

Figure 6. Reduction in cilia beats per minute from hamsters exposed to metals for two hours.

macrophages. Either extracellular or cellular mechanisms, or perhaps both, may be responsible for the observed decrease in phagocytosis.

In addition to this effect, both gaseous and particulate pollutants can also diminish selected hydrolytic enzyme activity of the alveolar macrophage either after *in vivo* or *in vitro* exposure to the pollutant.[34,44,46] These enzymes are a vitally important mechanism in the degradation of phagocytosed matter.

While defects in the phagocytic ability of the alveolar macrophage have been reported following pollutant exposure, there is little data on the actual site of the pollutant-induced damage. Since the initial interaction between the pollutant and the macrophage must occur at the plasma membrane of the cell, it is feasible that pollutant-induced lesion at this site could result in phagocytic dysfunction. Plasma membrane receptors are believed to be active in maintaining cellular integrity and function.

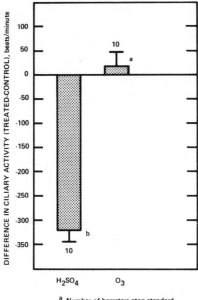

a Number of hamsters atop standard
error bars.

b Significantly different from
zero (p<.05).

Figure 7. Reduction in ciliary activity following a two-hour exposure to 900 $\mu g/m^3$ of H_2SO_4. Ozone (0.1 ppm x 3 hr) failed to alter the ciliary beats per minute as compared to controls.

Hadley *et al.*[47,48] have shown that pollutants can affect the membrane of the alveolar macrophage. Thirty-minute *in vitro* exposures to $NiCl_2$ ($> 1 \times 10^{-4}M$) or $CdCl_2$ ($> 2.2 \times 10^{-5}M$) resulted in significant dose-related decreases in the binding of antibody-sensitized sheep erythrocytes to the Fc receptor, as measured by a rosette assay (Figure 9). Such an effect could be expected to result in decreased phagocytosis by opsonized bacteria. Gases were also able to perturb the membrane. Lectin-treated alveolar macrophages from rabbits exposed to 980 $\mu g/m^3$ (0.5 ppm) O_3 or 13.167 $\mu g/m^3$ (7 ppm) NO_2 for three hours resulted in an increased rosetting of untreated autologous erythrocytes, possibly indicating the partial loss of the recognitive ability of these cells.

When this defense system is ineffective or overwhelmed by environmental factors, the deposited microorganisms have the opportunity to multiply and overcome local antibacterial defenses. Bacterial clearance and growth curves, determined by microbiological plate counts, indicate that the enhanced pneumonia due to pollutant exposure (O_3, NO_2, Cd, Ni) appears

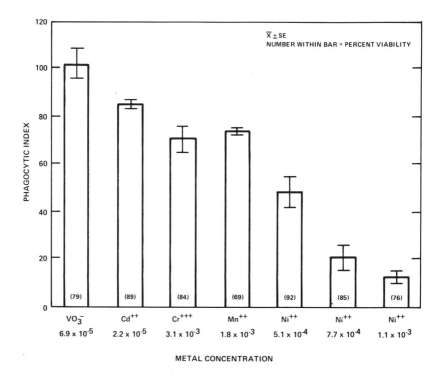

Figure 8. Influence of vanadium, cadmium, manganese, chromium and nickel on the phagocytic activity of pulmonary macrophage. Concentration expressed as moles of metal.

to be related to the ability of the inhaled microbes to multiply in the pollutant-exposed lung rather than to affect the general pulmonary clearance. The bacteria then invade the blood and death ensues.[13] A detailed description of the morphogenesis of the laboratory-induced infection has been given by Gardner et al.[49]

In these studies, control animals are capable of rapidly clearing the deposited streptococci; however, with the inhalation of an environmental pollutant, there is an inhibition of lung clearance, permitting those microbes with pathogenic potential to multiply and produce disease. It is of special interest that if a nonpathogenic type of bacteria such as *Serratia marcescens* is used, the clearance rates of the pollutant-exposed lung and the normal lung do differ (Figure 10).[13] When the time course of the infection is examined, it becomes evident that when the number of viable streptococci reach a critical level (approximately 10^5) within the lung, invasion of the blood occurs.[16]

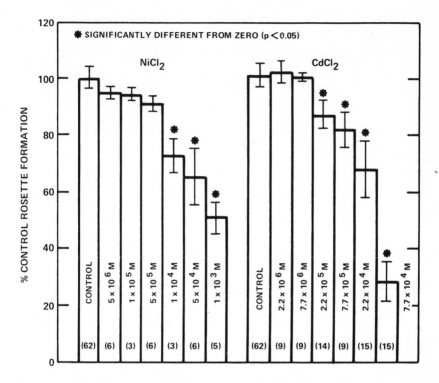

Figure 9. Effect of metals on antibody-mediated rosette formation by rabbit alveolar macrophages. Isolated cells were preincubated with either $NiCl_2$ or $CdCl_2$ at concentrations indicated for 30 minutes at $37°C$ prior to rosette assay. Bars indicate standard error.

DISCUSSION AND CONCLUSIONS

In assessing the role and usefulness of the infectivity model as a toxicological probe, it appears that it is a sensitive indicator of biological effects. The central feature of this whole animal infectivity model is that it permits observation of all the functioning available mechanisms of the host defense system against the growth of the inhaled microbes. The success or failure of the host to defend itself largely depends on the state of its interlocking pulmonary defense mechanisms. The sensitivity of the model is most probably due to this fact, in that it reflects the net effect of a number of subtle alterations in a number of subcomponents of the pulmonary defense system, none of which can be observed separately to be affected at the same pollutant concentrations. The importance of the interaction between the host, the environmental chemical and the infectious agent

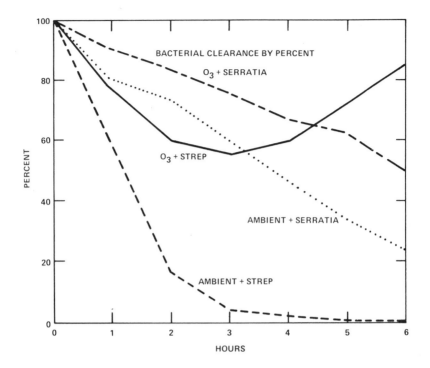

Figure 10. Comparison of the bacterial clearance for streptococci and *Serratia marcescens* from the lungs of mice after receiving a three-hour exposure to ozone and those controls that received clean air. With ozone exposure, the animals were capable of clearing the nonpathogen *Serratia* but were unable to cope with the Streptococci.

thus becomes increasingly clear. Such a model will provide a better understanding of how the lung protects itself against insults and should lead to a better understanding of the pathogenesis of respiratory diseases.

REFERENCES

1. American Thoracic Society. Medical Section of the National Tuberculosis and Respiratory Disease Association. *Am. Rev. Resp. Dis.* 101:116-150 (1970).
2. Rylander, R. "Pulmonary Defense Mechanisms to Airborne Bacteria," *Acta Physiol. Scand. Suppl.* 306:1-89 (1968).
3. Green, G.M., and E.H. Kass. The Influence of Bacterial Species on Pulmonary Resistance to Infection in Mice Subjected to Hypoxia, Cold Stress and Ethanolic Intoxication. *Brit. J. Pathol.* 46:360-366 (1965).
4. Witschi, H. "Exploitable Biochemical Approaches for the Evaluation of Toxic Lung Damage," *Essays in Toxicology,* Wayland J. Hayes, Ed. (New York: Academic Press).

5. Warshauer, D., E. Goldstein, P.D. Hoeprick, and W. Lippert. "Effect of Vitamin E and Ozone on the Pulmonary Antibacterial Defense Mechanism," *J. Lab. Clin. Med.* 83:228-240 (1974).

6. Mustafa, M.G., and S.D. Lee. "Pulmonary Biochemical Alterations Resulting from Ozone Exposure," *Ann. Occup. Hyg.* 19:17-26 (1977).

7. Menzel, D.B. "Toxicity of Ozone, Oxygen and Radiation." *Ann. Rev. Pharmacol.* 10:379-394 (1970).

8. Calabrese, E.J., W. Kajola and B.W. Carnow. "Ozone: A Possible Cause of Hemolytic Anemia in G-6-PD Deficient Individuals," *J. Toxicol. Environ. Health*, In Press.

9. Gardner, D.E., J.W. Illing, F.J. Miller and D.L. Coffin. "Enhancement of Effect of Exposure to Ozone and Nitrogen Dioxide by Exercise (Abst.) *Toxicol. Appl. Pharmacol.* 29:129 (1974).

10. Goldstein, E., G.W. Jordan, M.R. MacKenzie, and J.W. Osebold. "Methods for Evaluating the Toxicological Effects of Gaseous and Particulate Contaminants on Pulmonary Microbial Defense Systems," *Ann. Rev. Pharmacol. Toxicol.* 16:447-464 (1976).

11. Percora, D.V., and D. Yegian. "Bacteriology of the Lower Respiratory Tract in Healthy and Chronic Diseases," *New England J. Med.* 258:71-74 (1958).

12. Lauweryns, J.M., and J.H. Baert. "Alveolar Clearance and the Role of the Pulmonary Lymphatics," *Am. Rev. Resp. Dis.* 115:625-683 (1977).

13. Coffin, D.L. and D.E. Gardner. "Interaction of Biological Agents and Chemical Air Pollutants," *Ann. Occup. Hyg.* 15:219-234 (1972).

14. Green, G.M. "Lung Defense Mechanisms," *Med. Clin. N. Am.* 57:547-562 (1973).

15. Kass, E.H., G.M. Green, and E. Goldstein. "Mechanisms of Antibacterial Action in the Respiratory System," *Bacteriol. Rev.* 30:488-497 (1966).

16. Gardner, D.E., F.J. Miller, J.W. Illing, and J.M. Kirtz. "Alterations in Bacterial Defense Mechanisms of the Lung Induced by Inhalation of Cadmium," *Bull. Europ. Physiol. Resp.* 13:157-174 (1977).

17. Williams, D.M., J.A. Krick, and J.S. Remington. "Pulmonary Infection in the Compromised Host," *Am. Rev. Resp. Dis.* 114:359-394 (1976).

18. Graham, J.A., and D.E. Gardner. "Effects of Metals on Pulmonary Defense Mechanisms Against Infectious Disease." *Proc. Conf. on Environ. Toxicol.,* Wright-Patterson Air Force Base, Dayton, OH (1977).

19. Miller, F.J., J.W. Illing, and D.E. Gardner. "Effect of Urban Ozone Levels on Laboratory Induced Respiratory Infections," *Toxicol. Lett.* 2:163-169. (1978).

20. Henry, M.C., J. Findlay, J. Spangler, and R. Ehrlich. "Chronic Toxicity of NO_2 in Squirrel Monkeys," *Arch. Environ. Health* 20:566-570 (1970).

21. Enrlich, R. "Effect of Nitrogen Dioxide on Resistance to Respiratory Infection," *Bact. Rev.* 30:604-614 (1966).

22. Fairchild, G.A., J. Roan, and J. McCarroll, "Effect of Sulfur Dioxide on the Pathogenesis of Murine Influenza Infection," *Arch. Environ. Health* 25:174-182 (1972).

23. Gardner, D.E., D.L. Coffin, F.J. Miller, and E.J. Blommer. "Relationship Between NO₂ Concentration, Time and Level of Effect Using an Animal Infectivity Model," *Proceed. Int. Conf. Photochemical Oxidant Poll. and Its Control,* EPA Pub. No. 600-3-77-001, Vol. 1 (1977), pp. 513-525.

24. Adkins, B., and D.E. Gardner. "Measurement of Specific Trace Metal Toxicity Using a Microbiological Infectivity Model System," *Proc. First Int. Cong. Toxicol.,* Toronto, Canada, (1977).

25. Ehrlich, R. "Interaction Between Nitrogen Dioxide and Respiratory Infection," Scientific Seminar on Automotive Pollutants, EPA Pub. No. 600-9-75-003. Vol. 1, 1-9 (1975).

26. Coffin, D.L., D.E. Gardner, and E.J. Blommer. "Time-Dose Response in an Infectivity Model System," *Environ. Health Pers.* 13:11-15 (1976).

27. Gardner, D.E., F.J. Miller, E.J. Blommer, and D.L. Coffin. "Influence of Exposure Mode on the Toxicity of NO₂," *Environ. Pers.* In Press.

28. Gardner, D.E., and J.A. Graham. "Increased Pulmonary Disease Mediated Through Altered Bacterial Defenses," *Proc. 16th Ann. Hanford Biol. Symp.,* September 27-29 (1976).

29. Gardner, D.E., F.J. Miller, J.W. Illing, and J.M. Kirtz. "Increased Infectivity with Exposure to Ozone and Sulfuric Acid," *Toxicol. Lett.* 1:59-64, (1977).

30. Ehrlich, R., J.C. Findlay, J.D. Fenters, and D.E. Gardner. "Health Effects of Short-term Exposure to NO₂ - O₃ Mixtures," *Environ. Res.* 14:223-231 (1977).

31. Fenters, J. "Effect of Sulfuric Acid and Carbon Particle Mixtures on Bacterial and Viral Pneumonia in Mice," *Proc. Catalyst Res. Program Sulfuric Acid Res. Rev. Conf.* Hendersonville, NC. January 31 - February e, (1977). (EPA sponsored).

32. Adkins, B., and D.E. Gardner. "Effect of Nickel on the Enhancement of Induced Streptococcal Infections," (Abst.) *Proc. Am. Soc. Microbiol.* B-45: 18 (1976).

33. Coffin, D.L., D.E. Gardner, and D.B. Menzel. "Influence of Metals on Susceptibility to Infection and Reactivity to Irritating Substances," *Proc. Conf. on Factors Influencing Metabolism and Toxicity of Metals,* Stockholm, Sweden, July 17-23 (1977).

34. Waters, M.D., D.E. Gardner, C. Aranyi, and D.L. Coffin. "Metal Toxicity for Rabbit Alveolar Macrophages *In Vitro,"* *Environ. Res.* 9:32-47, (1975).

35. Cook, J.A., E.O. Hoffman, and N.R. DiLuzio. "Influence of Lead and Cadmium on the Susceptibility of Rats to Bacterial Challenge," *Proc. Soc. Exp. Biol. Med.* 150:741-747 (1975).

36. Hemphill, F.E., M.L. Kaeberle, and W.B. Buck. "Lead Suppression of Mouse Resistance to *Salmonella Typhimurium,"* *Science.* 172:1031-

37. Graham, J.A., D.E. Gardner, M.D. Waters, and D.L. Coffin. "Effect of Trace Metals on Phagocytosis by Alveolar Macrophages," *Infect. Immunol.* 11:1278-1283 (1975).

38. Goldstein, E., W.S. Tyler, P.D. Heoprich, and C. Eagle. "Ozone and the Antibacterial Defense Mechanisms," *Arch. Environ. Health* 26: 202-204 (1973).

39. Gardner, D.E., R.S. Holzman, and D.L. Coffin. "Effect of Nitrogen Dioxide on Pulmonary Cell Population," *J. Bactenol.* 98:1041-1043 (1969).

40. Adalis, D., D.E. Gardner, F.J. Miller, and D.L. Coffin. "Toxic Effects of Cadmium on Ciliary Activity Using a Tracheal Ring Model System," *Environ. Res.* 13:111-120 (1977).

41. Adalis, D., D. E. Gardner, and F. J. Miller. "Cytotoxic Effects of Nickel on Hamster Tracheal Rings," Am. Rev. Resp. Dis. 118:347-354 (1978).

42. Dalhamn, T. "Studies on Tracheal Ciliary Activity," *Am. Rev. Resp. Dis.* 89:870-877 (1964).
Healthy Rats and Rats Exposed to Irritant Gases (SO$_2$, H$_3$N, HCHO) *Acta Physiol. Scand. Suppl.* 36:1 (1956).

44. Waters, M.D., D.E. Gardner, and D.L. Coffin. "Cytotoxic Effects of Vanadium on Rabbit Alveolar Macrophages *in vitro.*" *Toxicol. Appl. Pharmacol.* 28:253-263 (1974).

45. Coffin, D.L., D.E. Gardner, R.S. Holzman, and F.J. Wolock. "Influence of Ozone on Pulmonary Cells," *Arch. Environ. Health* 16:633-636 (1968).

46. Hurst, D.J., D.E. Gardner, and D.L. Coffin. "Effect of Ozone on Acid Hydrolases of the Pulmonary Alveolar Macrophages," *J. Reticuloendothel. Soc.* 8:288-300 (1970).

47. Hadley, J.G., D.E. Gardner, D.L. Coffin, and D.B. Menzel. "Inhibition of Antibody Mediated Rosette Formation by Alveolar Macrophages," *J. Reticuloendothel, Soc.* 22(5):417-425,(1977).

48. Hadley, J.G., D.E. Gardner, D.L. Coffin, and D.B. Menzel. "Effect of Ozone and Nitrogen Dioxide Exposure of Rabbits on Binding of Autologous Red Cells to Alveolar Macrophages," *Int. Conf. on Photochem. Oxidants and Its Control,* EPA 600-3-77-001, Vol. 1 (1977), pp. 505-512.

49. Gardner, D.E., C.L. Sanders, I.G. Hadley, and R.R. Adee. "Morphogenesis of streptococcal Pneumonia," abstract presented at the American Society for Pharmacology and Experimental Therapeutics, Houston, TX, 1978.

CHAPTER 7

BIOLOGICAL EFFECTS OF
ENVIRONMENTAL POLLUTANTS: METHODS
FOR ASSESSING BIOCHEMICAL CHANGES

M. G. Mustafa

Department of Medicine, School of Medicine, and
Division of Environmental & Nutritional Sciences,
School of Public Health
University of California
Los Angeles, California   90024

S. D. Lee

The Health Effects Research Laboratory
Environmental Protection Agency
Cincinnati, Ohio   45268

INTRODUCTION

Biochemical changes may represent relatively more sensitive and early
indices of pollutant damage than the conventional pathophysiological
changes. We have studied the biological effects of ozone, nitrogen dioxide
and oxides of sulfur in terms of biochemical changes in target organs
after exposure of animals to one of these pollutants. For ozone and nitrogen dioxide the target organ appears to be the lung; for oxides of sulfur
and, to some extent, nitrogen dioxide the effects seem to reach the extrapulmonary tissues as well. A series of biochemical parameters, including
many aspects of intermediary metabolism, have been examined. Biochemical studies have been carried out at the level of whole animals,
organs, tissues, cells and subcellular organelles. The studies have enabled
us to define the dose-response relationship, the threshold limit value and
the reversible or irreversible nature of pollutant effects. The biochemical
data have been useful in correlating morphological and physiological
changes in target organs.

There is growing concern over the potential health effects of environmental contaminants. As more and more the identities of these conatminants are being known, their possible relationship with a multitude of health effects is becoming apparent. Environmental contaminants may find access to our body through food, drink and the air we breathe. This chapter will be confined to contaminants that pollute the breathing air and interact with the lung.

Although from the epidemiological point of view the observed health effects may be referred to a single pollutant or a group of pollutants, a considerable gap remains as to the precise relationship between the pollutant(s) and their effects, the mode of action of the pollutant(s) and the rate at which the health effects occur until the epidemiological endpoints are reached. Animal studies have generally been conducted to determine pollutant effects in terms of the degree and mode of occurrence. Again, relatively high-level, short-term effects of pollutants have been amply studied, which may be appropriate from the standpoint of inhalation toxicology or environmental toxicology. However, the relevance of these studies with respect to conditions in real life, *i.e.,* a long-term exposure to low-level pollutants, has remained questionable. Nevertheless, the methodology for assessing pollutant effects developed during the course of these studies remains valuable for determining the low-level, long-term effects. Besides, the effects of high-level exposure provide the assurance that experimental animals are sensitive to the pollutant(s) in question.

Various pathophysiological or pathomorphological alterations have generally been considered the primary criteria for detecting and quantitating diseased conditions. However, it is now recognized that biochemical determinations offer valuable means of assessing functional changes before the pathological endpoints are reached. The same applies to the field of air pollution, where the lung is the target organ for pollutant interaction. The biochemical processes in conjunction with morphological changes in the lung possibly represent the most sensitive and relatively early events of pollutant damage.[1-3] Thus, it is important that pollutant effects be determined and interpreted in biochemical terms. Such an approach would lead to delineation of mechanisms of pollutant action and offer means of preventing the propagation of adverse effects.

The major objectives of this chapter are to discuss:

1. the merits and usefulness of various biochemical methods applied to determine the pulmonary effects of selected air pollutants, particularly their short- and long-term effects at ambient or near-ambient concentrations found in the polluted areas;
2. the cellular processes or biochemical mechanisms associated with pollutant effects in animals and their possible implications for humans;

3. relative sensitivity and reliability or reproducibility of various methods in evaluating pollutant effects;

4. comparison and interpretation of data obtained by various methods to understand the nature of short- and long-term effects and predict their occurrence in humans;

5. the need for more techniques for sensitive and accurate determination of pollutant damage, especially development of biochemical screening techniques suitable for humans; and

6. reversibility versus irreversibility of pollutant effects and nutritional or other measures to prevent or lessen such effects.

BIOCHEMICAL STUDIES

Concerning the lung, evaluations of the toxic effects of environmental pollutants have commonly been made in terms of pulmonary function tests and pathomorphological lesions. More recently, the biochemical evaluation has received much attention, not only because biochemical changes are sensitive indicators but also because they reveal the mechanisms of toxicity and offer an improved understanding of structure-function relationship.[3-13]

The biochemical studies conducted in our laboratory include:

1. analytical and radiotracer methods, and tissue preparations (isolated lung, tissue slices and cellular and subcellular fractions) as applied to determining the metabolic changes in the lung;

2. determination of the nature and degree of metabolic changes *in vivo* and *in vitro* as indices for pollutant effects;

3. determination of duration and reversibility of pollutant effects; and

4. determination of influence of animal age and nutrition on pollutant effects.

From the investigation of metabolic and functional changes in animals at the level of cellular and subcellular organizations, we have been able to define the target organs for pollutant interaction. For the oxidant components of photochemical smog (ozone and nitrogen dioxide), the lung appears to be the target organ. For oxides of sulfur and, to some extent, nitrogen dioxide the effects seem to reach the blood and other extrapulmonary tissues.

The biochemical approaches made thus far are based on the known physical and chemical properties of ozone, nitrogen dioxide and oxides of sulfur. A series of biochemical parameters representing the metabolic profile have been examined after exposure of animals to a given pollutant. In general, these parameters exhibit stable baselines in tissues of control animals, and reproducible and dose-dependent changes in tissues of exposed animals. A systematic measurement of these parameters has led to an overall picture of the biochemical alterations in the target organs during and after a pollutant exposure.

In the following presentation, various parameters employed in the evaluation of pollutant effects will be discussed.

Body Weight, Lung Weight, and Protein and DNA Content

The body weight of animals serves as an overall index of growth. Growth of exposed animals relative to control animals may be affected by lack of food and water consumption over a period of days or weeks. This may be particularly true for relatively high-level exposures. Likewise, the lung weight of animals is essential. Both low- and high-level exposures are found to cause significant, often overwhelming, changes in lung weight. Pulmonary edema, either interstitial, alveolar or both, hemmorrhage, and infiltration of circulatory cells are the major causes of increase in lung weight during acute, high-level exposure, and sometimes at the early phase of low-level exposure. Such conditions, in which the lung weight may be doubled or tripled, reflect a gross injury to the lung due to pollutant toxicity. Increased lung weight is also observed following a low-level exposure (Table I). Such a gain, which takes place over a few days, reflects an increase in cellular mass of the lung, including the infiltrated circulatory cells. Initial injury and the subsequent repair and/or proliferative processes are thought to be responsible for the increase in lung weight. Conventionally, the alteration of lung weight is expressed as the ratio of dry weight to wet weight. In another way, the percent lung:body ratio is expressed as an index of alterations in lung weight.

Table I. Effect of 0.8 ppm Ozone Exposure on the
Lung Weight to Body Weight Ratio in Rats

Days of Exposure	Lung Weight/Body Weight Ratio (%)		Increase (%)
	Control	Exposed	
4	0.521	0.860	65
7	0.526	0.927	76

With the increase in lung weight, the protein content of the lung is altered. In acute, high-level exposure pulmonary edema, hemorrhage and inflammatory cells cause an increase of protein content. In low-level exposure the increase in protein content is primarily due to increased synthesis in the lung, although inflammatory cells are thought to contribute in part.

DNA content per cell is known to be constant. Thus, in acute exposure when the lung weight increases due to edema, the DNA content per mg of lung protein declines. After low-level exposure the DNA content is found to increase, reflecting cellular proliferation in the lung. Again, a part of the increase in DNA content may result from inflammatory cells.

From the foregoing discussions it is apparent that body and lung weight, and protein and DNA contents are important parameters that may serve as indices of pollutant toxicity. In various tissues, protein and DNA contents are useful in expressing enzyme activities. In lung tissue the expression of enzyme activities per mg of protein or DNA may not be considered valid, as the absolute values of these two entities are not constant. The best means of expressing enzyme activities is in units per lung. Expression per mg of dry lung is a conventional method, but the solutes of edema fluid may contribute to the dry weight of the lung.

Oxidative and Energy Metabolism

The major sites for cellular oxidative and energy metabolism are located in the mitochondria. Alterations of mitochondrial functions are sensitive indicators of cell damage, inasmuch as disruption of mitochondrial function may be rate-limiting to cell survival following exposure to certain pollutants. Mitochondrial function can be determined in lung homogenate and isolated mitochondria. The procedure for isolation of lung mitochondria by differential centrifugation has been reported.[8,9,14,15] Most commonly, the respiratory activity of mitochondria is assayed by a polarographic or manometric measurement of oxygen consumption in the presence of a substrate, *i.e.,* 2-oxoglutarate, succinate, glycerol-1-phosphate and ascorbate-Wurster's blue. Characteristics of lung mitochondrial substrate utilization have been reported.[8,9,14,15] Effects of ozone and nitrogen dioxide on mitochondrial oxygen consumption have been amply documented using an *in vitro* system, *i.e.,* directly exposing tissue homogenate or isolated mitochondria to one of these gases.[8,16,17] In general, these oxidants depress the respiratory activity, possibly by altering the mitochondrial membrane structure or inhibiting various dehydrogenases.[8] Various sulfites and sulfates are also found to inhibit the *in vitro* respiration in tissue homogenate or mitochondria. However, the metal cations, *e.g.,* Cu^{++}, Zn^{++} and Cd^{++}, seem to be important determinants in the inhibition caused by these salts. Under *in vivo* conditions, the inhibition of lung mitochondrial oxygen consumption has been observed after short-term, high-level ozone exposures, such as 2-4 ppm ozone for 4-8 hr.[8,9] With relatively long-term, low-level exposures to ozone or nitrogen dioxide, there is a stimulation of mitochondrial oxygen consumption in lung tissue (Table II). The stimulation is explained on the basis of an increased mitochondrial population in lung tissue occurring in response to ozone injury.[11-13]

Table II. Effect of 0.8 ppm Ozone Exposure for
Four Days on Oxygen Consumption in Rat Lung Tissue

Tissue Preparation	O_2 Consumption (μl/hr lung)		Increase (%)
	Control	Exposed	
Tissue Slices[a]	1,170	1,530	31
Homogenate[b]	1,172	2,671	51

[a]In the presence of 2, 4-dinitrophenol.
[b]In the presence of succinate.

Other functions of mitochondria are oxidative phosphorylation, measured in terms of P:O ratio, and respiratory control. Studies involving these functions of lung mitochondria have also been reported.[8,14,15] These parameters are reproducible and their assays can be performed accurately. Other parameters of mitochondrial function that can be used in monitoring pollutant effects include the rates of ion uptake, swelling and contraction in isotonic media. Isotopic, spectrophotometric and light scattering techniques for these measurements have been well described.[18,19] These *in vitro* tests for isolated mitochondria can be correlated with mitochondrial morphology *in situ* of lung tissue in thin sections, *i.e.*, in electron micrographs.

In addition to functions of mitochondria, partial electron transport reactions in lung microsomes can be assayed to monitor pollutant effects. Activities of cytochrome *c* reductases pertaining to NADH and NADPH have been used as sensitive markers for microsomal membranes.[12,13] Preparation and properties of lung microsomes have been described by several workers.[13,20]

Glucose and Lipid Metabolism

Glucose metabolism, particularly consumption of glucose, production of pyruvate, lactate and carbon dioxide, and incorporation of glucose carbon into lipid and protein fractions of lung tissue, may offer an important set of screening tests for pollutant effects on the lung. The characteristics of glucose utilization in isolated perfused lung and tissue slices have been reported recently.[21] Any form of lung injury seems to greatly alter (actually stimulates) glucose consumption and lactate production. It is not clear why glucose consumption is stimulated. In acute lung injury, *e.g.*, due to high-level exposure to ozone or nitrogen dioxide, glucose consumed is virtually quantitatively converted to lactate. Under *in vivo* conditions most of the lactate produced is probably channeled to pulmonary circulation.[21] Glucose consumption is also augmented after low-level exposure to ozone or nitrogen dioxide (Table III). Under these conditions, approximately 50% of the glucose consumed may be converted to lactate, but the rest actually may be utilized as assessed from the $^{14}CO_2$ produced and ^{14}C from labeled glucose incorporated into protein and lipid fractions of lung tissue.

Table III. Effect of 0.8 ppm Ozone Exposure for Four Days on Glucose Metabolism in Rat Lung Tissue Slices

| | Metabolite (μmoles/hour · lung) | | Increase (%) |
	Control	Exposed	
Glucose (consumed)	31	47	58
Pyrovate (produced)	2.6	3.6	38
Lactate (produced)	15	22	47

In addition to the glycolytic pathway, glucose consumption via the hexose monophosphate shunt pathway is also augmented as a result of ozone or nitrogen dioxide exposure.[4-6,12,13] The augmentation of this pathway is inferred from the increased activities of glucose-6-phosphate dehydrogenase (G6PD) and 6-phosphogluconate dehydrogenase (6PGD), and from the elevated ratio of CO_2 from 1-^{14}C-glucose to CO_2 from 6-^{14}C-glucose.[21-23] Table IV shows low-level ozone effects on G6PD and 6PGD activities. Lipid metabolism, particularly the rate of utilization of various substrates for lipid synthesis and the fatty acid profile, offers important parameters for detection of pollutant effects on the lung. Lipid metabolism has been extensively studied in the lung,[24-26] and there are several reports on the effects of ozone and nitrogen dioxide on the alteration of fatty acid composition.[27-29]

Table IV. Effect of 0.8 ppm Ozone Exposure for Seven Days on
G6PD and 6PGD Activities in Rat Lung Cytosol

Enzyme	Activities (μmol NADPH formed/min · lung)		Increase (%)
	Control	Exposed	
G6PD	2.96	5.11	73
6PGD	1.76	2.82	60

Protein and Nucleic Acid Synthesis

Various parameters involving protein and nucleic acid metabolism are frequently examined in toxicological studies. Alterations of proteins and nucleic acid synthesis often serve toward unfolding the mechanisms of toxicity due to various environmental agents.

The enhancement of enzymatic activities in the lung observed after low-level oxidant exposure or during recovery from a high-level exposure may be due to increased synthesis of proteins and enzymes, and/or activation of enzymes.[13] In most studies with the lung, the alterations of enzymes or protein synthesis have been demonstrated using the method of labeled amino acid incorporation into proteins.[30,31] An example is given in Table V. Likewise, the alterations in nucleic acid (RNA and DNA) have been studied using labeled percursors (thymidine for DNA and uridine for RNA).[31]

Table V. Effect of 0.8 ppm Ozone Exposure on
[14]C-Leucine Incorporation into Lung Proteins *IN Vitro*

Days of Exposure	Rate of Incorporation (%) of control)
2	140
4	178
7	182

Metabolism of Sulfhydryls and Reducing Substances

Soluble sulfhydryl compounds, also called nonprotein sulfhydryls (NPSH), particularly reduced glutathione (GSH), constitute a major pool of reducing substances in the cell. A rat lung (weighing approximately 1 g) contains approximately 1.6 μmol of NSPH, up to 90% of which may be GSH. Both GSH and NPSH can be assayed conveniently.[32,33] Their levels in lung tissue are altered in response to oxidant exposure.[4,12,13,33] In addition to GSH or NPSH, the SH groups of enzymes and proteins (PSH) can be assayed, and their levels are also altered as a result of oxidant exposure.[4,33] In high-level ozone exposure, e.g., 2-4 ppm for 4-8 hr, a significant depression of SH levels (GSH, NPSH and PSH) has been observed.[4,33] Under these exposure conditions mixed disulfides (PSSG) were shown to be formed between GSH and PSH.[33] Thus, assay of the level of GSSG offers another important parameter for detecting oxidant injury in the lung.

In contrast to acute, high-level exposure, which decreases the level of SH groups in lung tissue, low-level exposure to ozone or nitrogen dioxide results in an increase of GSH or NPSH level (Table VI). The level of SH groups, which have important cellular functions,[34,35] offers a sensitive indicator of lung tissue injury and repair during oxidant pollutant exposure.

Table VI. Effect of 0.8 ppm Ozone Exposure for Three Days on the Level of Soluble Sulfhydryls in Rat Lung Tissue

	Concentrations (μmol/lung)		Increase (%)
	Control	Exposed	
NPSH	1.6	2.2	38
GSH	1.3	1.7	31

Activities of several marker enzymes, such as glutathione reductase, disulfide reductase and glutathione-disulfide transhydrogenase, which involve sulfhydryl metabolism, can be assayed.[13,36] Activities of these enzymes show marked alterations as a result of ozone or nitrogen dioxide exposure (Table VII) and may be employed to monitor pollutant damage to the lung.

Some of the other reducing substances in the cell are NADH and NADPH. These metabolites, and also GSH, participate in many enzymatic functions that are critical to cellular viability, growth and reproduction. The levels of NADH and NADPH in lung tissue may serve as sensitive parameters for detecting oxidant pollutant effects. Activities of enzymes such as glucose-6-phosphate dehydrogenase, 6-phosphogluconate dehydrogenase, isocitrate dehydrogenase and NADH-NADP$^+$ transhydrogenases can be assayed in lung tissue to determine pollutant effects.

Table VII. Effect of 0.8 ppm Ozone Exposure for Seven Days on the Activities of Sulfhydryl Metabolizing Enzymes in Rat Lung Cytosol

Enzyme	Activities (μmol NADPH Oxidized/min · lung)		Increase (%)
	Control	Exposed	
Glutathione Reductase	1.8	2.37	30
Disulfide Reductase	1.68	2.52	50
Glutathione-Disulfide Transhydrogenase	0.95	1.38	45

Metabolism Related to Cellular Detoxification

Toxicity of oxidant pollutants is thought to involve free radical chain reactions and peroxidation of unsaturated lipids. Various toxic intermediates, *i.e.,* free radicals, superoxide anion, hydroxyl radical, hydrogen peroxide and lipid peroxides, generated in lung tissue during ozone or nitrogen dioxide exposure are possibly eliminated via enzymatic detoxification mechanisms. Activities of such enzymes as glutathione peroxidase, superoxide dismutase, peroxidase and catalase are associated with such mechanisms and are found to respond to oxidant exposure.[5,6,13] The activities of these enzymes serve, therefore, as sensitive indicators of lung tissue injury due to ozone or nitrogen dioxide exposure (Table VIII).

Table VIII. Effect of 0.8 ppm Ozone Exposure for Seven Days on Rat Lung Enzyme Activities Related to Antioxidant Protection

Enzyme	Activities (units/min · lung)[a]		Increase (%)
	Control	Exposed	
In Mitochondria			
Glutathione peroxidase	1.87	2.81	50
Superoxide dismutase	130	189	46
In Cytosol			
Glutathione peroxidase	0.98	1.37	39
Superoxidase dismutase	388	540	39

[a] Units are μmoles NADPH reduced for glutathione peroxidase and percent inhibition of cytochrome *c* reductase for superoxide dismutase.

Analyses of Reaction Products and
Metabolites After Exposure

In addition to measurements of antioxidant compounds (*e.g.,* GSH, NADH and NADPH) and enzymatic activities, analyses of certain reaction products and metabolites during or after an exposure may offer important tests for pollutant interaction with lung. In exposures involving ozone or nitrogen dioxide, two lines of approach have been considered or attempted. Detection and quantitation of free radicals or lipid peroxidation products in tissues may be a direct demonstration of these processes occurring in oxidant exposure. Reports of success in this direction,[5,6,37] however, have met with controversies.[13] In another approach, the metabolites of vitamins E and C may be quantitated to assess oxidant stress. It is thought that during exposure to ozone or nitrogen dioxide there is increased oxidative consumption of antioxidant substances, such as vitamins E and C. The sequential oxidation of vitamin E is thought to be:

$$D\text{-}\alpha\text{-tocopherol} \rightarrow D\text{-}\alpha\text{-tocopherylquinone} \rightarrow$$

$$D\text{-}\alpha\text{-tocopheronolactone} \rightarrow D\text{-}\alpha\text{-tocopheronic acid}$$

A measurement of the levels of the initial, intermediate and final metabolites in lung tissue, plasma and urine may serve as important tests for the degree of oxidant stress. Likewise, the levels of vitamin C (ascorbate) and its oxidation product dehydroascorbate can be monitored to evaluate the effects of oxidant.

For exposure involving sulfur dioxide, the levels of sulfite, sulfate, S-sulfonate and total sulfur content can be determined in lung tissue, plasma and urine. These tests will document whether the body burden of sulfur increases due to sulfur dioxide exposure.

Influences of Animal Age and Nutrition on
Metabolic Response to Pollutant Exposure

Exposure of animals to ozone or nitrogen dioxide results in metabolic changes in the lung. The magnitude of metabolic changes, which are essentially a response to injury, depends on the degree of exposure. However, at least two other factors have been found to influence the metabolic changes during a given oxidant exposure. These are the age and nutritional status of the animals. As has been successfully studied in rats, adult animals (2-3 months of age) are more susceptible to oxidant injury than young animals (1 month old) (Table IX); animals of age 6 months or older are the most susceptible. Likewise, animals poorly supplied with vitamin E exhibit much greater sensitivity to ozone or nitrogen dioxide exposure compared to animals receiving nutritional or supplemented dosage of vitamin E (Table X).[5,6,11] Vitamin C, GSH and other antioxidant compounds have been shown to be protective against oxidant toxicity.[39-42] Thus, in selecting animal models for studies of pollutant effects, the age and nutritional status must be considered to demonstrate reproducible and dose-dependent effects of pollutants.

Table IX. Influence of Animal Age on the Metabolic
Response of Rat Lungs to 0.8 ppm Ozone Exposure for Three Days

Animal Age (days)	QO_2 in Tissue Slices ($\mu l/hr \cdot lung$)[a]		Increase (%)
	Control	Exposed	
30	1,196	1,204	1
45	1,338	1,773	32
60	1,332	1,836	38
90	2,012	3,016	50

[a]In the presence of 2, 4-dinitrophenol.

Table X. Influence of Dietary Vitamin E Levels on the Metabolic
Response of Rat Lungs to 0.8 ppm Ozone Exposure for Four Days

Enzyme	Activity (% of control) at Vitamin E Level	
	10 ppm	50 ppm
Succinate Oxidase[a]	250	161
Glucose-6-Phosphate Dehydrogenase[b]	272	200
Glutathione Reductase[c]	164	147

[a]Control value, 1600 μl O_2/hour $\cdot$ lung
[b]Control value, 2.75 μmoles NADPH formed/min $\cdot$ lung
[c]Control value, 2.2 μmoles NADPH oxidized/min $\cdot$ lung

DISCUSSION AND CONCLUSIONS

We have considered how biochemical approaches can be applied to detect and quantitate the pulmonary effects of environmental pollutants. Many of the methods discussed are relatively simple and suitable for both lung tissue and extrapulmonary tissues, so that examination of pollutant effects can be made in latter tissues as well. It should be pointed out that although the biochemical methods provide considerable sensitivity and convenience in detecting and quantitating lung injury, they are not free of difficulties. The major disadvantage is that the biochemical changes represent an average of the pollutant effects on many cell types present in the lung.[43] Specific or critical changes in a small number of cells may not be apparent from biochemical measurements made in lung homogenate or tissue preparations involving the whole lung. Also, all biochemical parameters may not respond in equal degrees to a given pollutant exposure. To circumvent a part of these difficulties, it is important to examine a series of biochemical parameters and make a plot of the relative degree of their changes. The cause-and-effect relationship can be found in this way. Various experimental conditions that modulate pollutant effects should be applied, particularly to develop a sensitive model or models for the screening purpose. Furthermore, biochemical changes should be correlated with morphological changes to include specific or critical response of certain lung cells to a given pollutant. Several laboratories have documented that there is a good correlation between biochemical and morphological changes in the lung, particularly during the early phases of ozone or nitrogen dioxide exposures.[7,12,13,44-46] Thus, several lines of approach made in concert will not only provide valid detection and assessment data but also offer sensitive criteria for establishing threshold limit values for various environmental pollutants.

In conclusion, this presentation has pointed out that a series of biochemical parameters can be assayed in lung tissue, plasma and urine to determine the effects of pollutants, *e.g.*, ozone, nitrogen dioxide and sulfur dioxide, in experimental animals. These determinations will also permit the documentation of sequential biochemical processes and mechanisms of toxicity associated with pollutant exposure. Furthermore, a good many of these tests, which are noninvasive, can be developed into biomedical screening tests applicable to human population, particularly for the smog-polluted areas.

ACKNOWLEDGMENTS

This study was supported in part by USEPA contract 68-03-2221, SCE contract U2277914, USPHS-NIH grant HL-17719 and NIH-RCD award HL-00301 (to M. G. Mustafa).

The authors wish to thank Dr. E. J. Faeder of Southern California Edison Company and Dr. W. S. Simmons of Greenfield, Attaway & Tyler,

118

Inc. for their encouraging suggestions in developing the section on analyses of metabolites. Skilled technical assistance of Mr. Nabil Elsayed is acknowledged.

REFERENCES

1. Cross, C. E., A. J. De Lucia, A. K. Reddy, M. Z. Hussain, C. K. Chow, and M. G. Mustafa. "Ozone Interaction with Lung Tissue: Biochemical Approaches," *Am. J. Med.* 60:929 (1976).
2. Witschi, H. "Exploitable Biochemical Approaches for the Evaluation of Toxic Lung Damage," in *Essays in Toxicology*, W. J. Hayes, Ed., Vol. 6 (New York: Academic Press, 1975), p. 125.
3. Witschi, H., and M. G. Coté. "Biochemical Pathology of Lung Damage Produced by Chemicals," *Fed. Proc.* 35:89 (1976).
4. De Lucia, A. J., P. M. Hoque, M. G. Mustafa, and C. E. Cross. "Ozone Interaction With Rodent Lung. I. Effects on Sulfhydryls and Sulfhydryl-Containing Enzyme Activities," *J. Lab. Clin. Med.* 80:559 (1972).
5. Chow, C. K., and A. L. Tappel. "An Enzymatic Protective Mechanism Against Lipid Peroxidation Damage to Lungs of Ozone-Exposed Rats," *Lipids* 7:518 (1972).
6. Chow, C. K., and A. L. Tappel. "Activities of Pentose Shunt and Glycolytic Enzymes in Lungs of Ozone-Exposed Rats," *Arch. Environ. Health* 26:205 (1973).
7. Mustafa, M. G., A. J. De Lucia, G. K. York, C. Arth, and C. E. Cross. "Ozone Interaction with Rodent Lung. II. Effects on Oxygen Consumption of Mitochondria," *J. Lab. Clin. Med.* 82:357 (1973).
8. Mustafa, M. G., and C. E. Cross. "Effects of Short-Term Ozone Exposure on Lung Mitochondrial Oxidative and Energy Metabolism," *Arch. Biochem. Biophys.* 162:585 (1974).
9. Mustafa, M. G. "Augmentation of Mitochondrial Oxidative Metabolism in Lung Tissue During Recovery of Animals from Acute Ozone Exposure," *Arch. Biochem. Biophys.* 165:531 (1974).
10. Werthamer, S., P. D. Penha, and L. Amaral. "Pulmonary Lesions Induced by Chronic Exposure to Ozone," *Arch. Environ. Health* 29:164 (1974).
11. Mustafa, M. G. "Influence of Dietary Vitamin E on Lung Cellular Sensitivity to Ozone in Rats," *Nutr. Rep. Int.* 11:473 (1975).
12. Mustafa, M. G., and S. D. Lee. "Pulmonary Biochemical Alterations Resulting from Ozone Exposure," *Ann. Occup. Hyg.* 19:17 (1976).
13. Mustafa, M. G., A. D. Hacker, J. J. Ospital, M. Z. Hussain, and S. D. Lee. "Biochemical Effects of Environmental Oxidant Pollutants in Animal Lungs," in *Biochemical Effects of Environmental Pollutants*, S. D. Lee, Ed. (Ann Arbor, MI: Ann Arbor Science Publishers, Inc., 1977), p. 59.
14. Reiss, O. K. "Studies of Lung Metabolism. I. Isolation and Properties of Subcellular Fractions from Rabbit Lung," *J. Cell. Biol.* 30:45 (1966).

15. Fisher, A. B., A. Scarpa, K. F. La Noue, D. Bassett, and J. R. Williamson. "Respiration of Rat Lung Mitochondria and the Influence of Ca^{++} on Substrate Utilization," *Biochemistry* 12:1438 (1973).

16. Ramazzotto, L., C. R. Jones, and F. Cornell. "Effect of Nitrogen Dioxide on the Activities of Cytochrome Oxidase and Succinic Dehydrogenase on Homogenates of Some Organs of the Rat *Life Sci.* 10:601 (1971).

17. Freebairn, H. T. "Reversal of Inhibitory Effects of O_3 and O_2 Uptake by Mitochondria," *Science* 126:303 (1957).

18. Packer, L. "Energy-Linked Low Amplitude Mitochondrial Swelling," *Methods Enzymol.* 10:685 (1967).

19. Hunter, E. E., Jr., and E. E. Smith. "Measurement of Mitochondrial Swelling and Shrinking—High Amplitude," *Methods Enzymol.* 10: 689 (1967).

20. Matsubara, T., and Y. Tochi o. "Electron Transport Systems of Lung Microsom and Their Physiological Functions. I. Intracellular Distribution of Oxidative Enzymes in Lung Cells," *J. Biochem.* 70: 981 (1971).

21. Tierney, D. F., and S. E. Levy. "Glucose Metabolism," in *The Biochemical Basis of Pulmonary Function*, R. G. Crystal, Ed. Vol. 2 (New York: Marcel Dekker, 1976), p. 105.

22. Katz, J., and H. G. Wood. "The Use of $^{14}CO_2$ Yield From Glucose-1- and -6-^{14}C for the Evaluation of the Pathways of Glucose Metabolism," *J. Biol. Chem.* 238:517 (1963).

23. O'Neil, J. J., and D. F. Tierney. "Rat Lung Metabolism: Glucose Utilization by Isolated Perfused Lungs and Tissue Slices," *Am. J. Physiol.* 226:867 (1974).

24. Tierney, D. F. "Lung Metabolism and Biochemistry," *Ann. Rev. Physiol.* 36:209 (1974).

25. Mason, R. J. "Lipid Metabolism," in *The Biochemical Basis of Pulmonary Function*, R. G. Crystal, Ed., Vol. 2 (New York: Marcel Dekker, 1976), p. 127.

26. Naimark, A. "Cellular Dynamics and Lipid Metabolism in the Lung," *Fed. Proc.* 32:1967 (1973).

27. Roehm, J. N., J. C. Hadley, and D. B. Menzel. "Antioxidant vs. Lung Diseases," *Arch. Int. Med.*, 128:88 (1971).

28. Roehm, J. N., J. C. Hadley, and D. B. Menzel. "The Influence of Vitamin E on the Lung Fatty Acids of Rats Exposed to Ozone," *Arch. Environ. Health* 24:237 (1972).

29. Shimasaki, H., T. Taketori, W. R. Anderson, H. L. Horten, and O. S. Privett. "Alteration of Lung Lipids in Ozone Exposed Rats," *Biochem. Biophys. Res. Commun.* 68:1256 (1976).

30. Massaro, D., H. Weiss, and M. R. Simon. "Protein Synthesis and Secretion by Lung," *Am. Rev. Resp. Dis.* 101:198 (1970).

31. Witschi, H. "Qualitative and Quantitative Aspects of the Ribonucleic Acid and Protein in the Liver and the Lung of the Syrian Golden Hamster," *Biochem. J.* 136:781 (1973).

32. Sedlak, J., and R. H. Lindsay. "Estimation of Total, Protein-Bound and Nonprotein Sulfhydryl Groups in Tissue with Ellman's Reagent," *Anal. Biochem.* 25:192 (1968).

33. De Lucia, A. J., M. G. Mustafa, M. Z. Hussain, and C. E. Cross. "Ozone Interaction with Rodent Lung. III. Oxidation of Reduced Glutathione and Formation of Mixed Disulfides Between Protein and Nonprotein Sulfhydryls," *J. Clin. Invet.* 55:794 (1975).

34. Jocelyn, P. C., Ed. *Biochemistry of the SH Groups,* (New York: Academic Press, 1972).

35. Kosewer, N. S., and E. M. Kosower. "The Glutathione-Glutathione Disulfide System," in *Free Radicals in Biology,* W. A. Pryor, Ed., Vol. II (New York: Academic Press, 1976), p. 55.

36. Tietze, F. "Disulfide Reduction in Rat Liver," *Arch. Biochem. Biophys.* 138:177 (1970).

37. Thomas, H. V., P. K. Moeller, and R. L. Lyman. "Lipoperoxidation of Lung Reports in Rats Exposed to Nitrogen Dioxide," *Science* 159:532 (1968).

38. Ostpial, J. J., A. D. Hacker, N. Elsayed, M. G. Mustafa, and S. D. Lee. "Influence of Age on the Effect of Ozone Exposure in Rat Lungs," *Am. Rev. Resp. Dis.* 115:235 (1977).

39. Stokinger, H. E. "Evaluation of the Hazards of Ozone and Oxides of Nitrogen," *AMA Arch. Health* 15:181 (1957).

40. Matzen, R. N. "Effect of Vitamin C and Hydrocortisone on the Pulmonary Edema Produced by Ozone in Mice," *J. Appl. Physiol.* 11:105 (1957).

41. Fukase, O., K. Isomura, and H. Watanabe. "Effect of Ozone on Vitamin C *In Vivo,*" *Taiki-osen Kenkyo* (Japanese) 10:13 (1975).

42. Fairchild, E. J., S. D. Murphy, and H. E. Stokinger. "Protection by Sulfur Compounds Against Air Pollutants Ozone and Nitrogen Dioxide," *Science* 130:861 (1959).

43. Sorokin, S. P. "Properties of Alveolar Cells and Tissues that Strengthen Alveolar Defenses," *Arch. Int. Med.* 126:450 (1970).

44. Evans, M. J., L. J. Cabral, R. J. Stephens, and G. Freeman. "Renewal of Alveolar Epithelium in the Rat Following Exposure to NO_2," *Am. J. Pathol.* 70:175 (1973).

45. Stephens, R. J., M. F. Sloan, M. J. Evans, and G. Freeman. "Early Response of Lung to Low Levels of Ozone," *Am. J. Pathol.* 74:31 (1974).

46. Schwartz, L. W., D. L. Dungworth, M. G. Mustafa, B. K. Tarkington, and W. S. Tyler. "Pulmonary Responses of Rats to Ambient Levels of Ozone. Effects of 7-day Intermittent of Continuous Exposure," *Lab. Invest.* 34:565 (1976).

CHAPTER 8

APPLICATIONS OF LUNG ORGAN CULTURE
IN ENVIRONMENTAL INVESTIGATIONS

Rajendra S. Bhatnagar, M. Zamirul Hussain
and John C. Belton*

Laboratory of Connective Tissue Biochemistry
School of Dentistry
University of California
San Francisco, California 94143

INTRODUCTION

The lungs are a particularly vulnerable target for airborne toxins because they present, among all parts of the body, the largest exposed surface— nearly 70 m^2.[1] Their susceptibility is enhanced by the forced circulation of the atmosphere through them as a part of their vital function. Elaborate defense mechanisms exist in the body to protect the lungs from a great variety of harmful agents such as infectious organisms. When exposure to noxious substances results in tissue damage, lungs respond by repairing and, to some extent, regenerating the injured tissues in the same manner as do injured tissues elsewhere in the body. With increased concentrations of atmospheric toxins arising from energy-generation, industrial and trans- portation-related exhausts as well as from agricultural and other procedures involving dispersal of chemical agents, the defensive and reparative- regenerative mechanisms of the lungs are stressed to their limits. Under these conditions, tissue injury, excessive repair and scarring lead to altered tissue architecture and impaired function. Unfortunately, our awareness that large populations may be unavoidably exposed to deleterious sub- stances in this manner has not been accompanied by the development of adequate techniques to rapidly and efficiently detect and determine the

*Department of Biology, California State University, Hayward, California.

sequence of events in pulmonary injury that eventually leads to impaired respiratory function.

The recognition that an almost unlimited number of old and new environmental toxins may initiate injury in the lungs which may be exacerbated to emphysematous disease or culminate in pulmonary fibrosis presents major challenges. It is necessary to identify the agents that may contribute to lung damage. This information is essential for developing regulatory guidelines in determining the limits of exposure and for the establishment of protective measures. It is of utmost importance to understand the fundamental processes underlying tissue injury initiated by the toxins since this information is necessary for the development of measures to prevent such damage and for the development of measures that would interfere with the pathological reaction in the continuous presence of the toxin. The evaluation of deleterious effects of toxins on lungs can be made by following physiological changes, such as alterations in pulmonary function, and by morphological examination of the lungs. Detectable functional and morphological changes represent late stages in the development of pulmonary disorder. While they are useful for identifying etiologic agents, they are not sensitive enough to assay early changes. Interactions of tissues with toxins result in chemical alterations in cell membranes, and other organelles and cellular metabolic patterns are significantly modified. Since chemical and biochemical changes are easily detected with a high degree of sensitivity, they provide a useful means of following the course of environmental injury.

Such changes can also be easily related to incipient structural alterations which may not become apparent by morphological examination and pulmonary function tests until the disease process has reached an advanced stage. Since the earliest signs of environmental injury may appear in lung cells, which have the greatest susceptibility, development of procedures to permit the evaluation of biochemical alterations in these cells and in their macromolecular products, both intra- and extracellular, would provide insight into the injury response of the lung.

BIOCHEMICAL CONSEQUENCES OF
ENVIRONMENTAL INJURY

Increasing concern over the quality of the atmosphere has led to a major surge in investigations into the biochemical effects of environment-tissue interaction. Although an overall picture has not emerged, recent investigations suggest that the earliest events in environmental toxin-induced injury may involve the direct interaction of the toxic material with cell membranes and other organelles. The genetic material as well as regulatory sites may be affected. The train of events thus set off results in changes in metabolic characteristics. Altered availability of energy and modified precursor pools combined with damaged regulatory sites result in drastically altered synthesis of intra- and extracellular macromolecules. The resultant injury evokes the body's repair mechanisms. Inflammation and immune responses are initiated and the architectural modifications leading to functional impairment are well on their way. Some of these events are summarized in Figure 1. These aspects of environmental injury have been reviewed extensively.[2-6] Recent evidence suggests that the metabolic patterns change as a function of exposure and, after prolonged exposures, the tissues may adapt to the toxic environment, with a concomitant readjustment of metabolic rates that may then appear to be normal.[7]

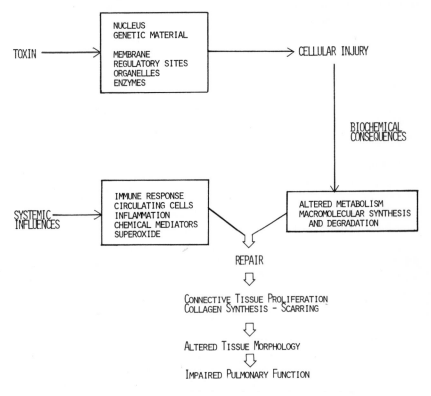

Figure 1. Toxin-tissue interaction leading to pulmonary dysfunction.

A major consequence of lung injury by a variety of environmental agents is the localized breakdown of connective tissue elements (emphysematous lesions) or the proliferation of connective tissue (fibrosis). The major extracellular macromolecules of connective tissue, collagen and elastin are the major structural elements of lungs. They are directly involved in the mechanical aspects of lung function. Collagen is the major constituent of alveolar membranes across which gaseous exchange occurs. Little is known about the fate of elastin during the development of atmospheric pollutant-induced lung disease; however, the rate of collagen synthesis is known to be enhanced markedly.[8-10] Examination of lungs of experimental animals exposed to toxic atmospheres revealed that after short-term exposures there was an increase in the content of collagen both by morphological[10,11] and biochemical criteria.[12,13] However, after longer exposures, lungs of animals that showed gross deterioration of pulmonary function and markedly altered lung morphology, did not appear to have significantly different collagen contents in comparison to control animal lungs.[14] Similar observations have been reported in fibrotic human lungs.[15-17]

These studies point to the importance of determining the earliest biochemical events in the initiation of pulmonary injury by environmental agents.

EXPERIMENTAL MODELS FOR EVALUATING PULMONARY TOXICITY

As summarized in Tables I and II, a variety of *in vivo* and *in vitro* procedures are used in studies on environmental toxicology. Our present concern over the quality of the atmosphere arose largely from epidemiological studies, which showed that human populations exposed to certain substances present in the air suffered deleterious effects, which could be ascribed to these substances. Unfortunately, such studies can detect only the more advanced stages of the environment-induced disease and do not provide sensitive procedures for elucidating the early biochemical events. Determination of biochemical changes often requires procedures not

Table I. Models for Evaluating Pulmonary Toxicity *in vivo* Studies

Epidemiology	Shows delayed effects, not sensitive to early changes, not suited for experimental studies.
Animal Models	Subject to infection; susceptibility to infection increased by exposure; biological variability; effect complicated by systemic influences (immune, inflammation, edema, nutrition); requires high dose of exposure, low sensitivity, complications due to adaptability.

Table II. *In Vitro* Models for Evaluating Pulmonary Toxicity

Cell Cultures	Very sensitive; allow measurements of effects on individual cell types and cellular processes, but intercellular and cell matrix interactions absent; loss of differentiated function.
Organ Cultures	Very sensitive; retain morphological and cellular integrity and cell-cell, cell-matrix interactions; remain differentiated; mimic *in vivo* response but not cluttered by systemic influences; rapidly reproducible and highly controllable.

feasible in humans. An alternative to human epidemiological studies is the use of experimental animals. Much of the current knowledge of the toxicological and biochemical interactions of environmental toxins with lungs and the resultant functional and structural changes is based on such studies. The animals are placed in inhalation chambers, in environments that can be modified to include the test toxins, and regulated. Such studies permit the evaluation of biochemical and other changes as a function of tissue since it is possible, by sacrifice, to remove the lungs from individual animals for different tissues. However, because of the inherent biological variability among individual animals, even from the same litter, it is sometimes difficult to interpret the data: for instance, individual animals may have different growth rates, susceptibility to bacterial and viral infection and may also differ in the effectiveness of their protective mechanisms.[18,19] Additional complications arise from the enhanced susceptibility of the toxin-exposed animals to infection. On a more direct level, the evaluation of biochemical changes in the lungs is rendered complex by various systemic influences, infiltrating cells, circulating proteins, immune complexes and other material, all of which converge at the site of injury as part of the body's defense and repair mechanisms.

Physiological and morphological studies have shown that the initial events in the induction of lung injury are parenchymal. The parenchyma comprises nearly 80% of the lung mass. The tissue consists of nearly 40 different types of cells embedded in specific arrangements within a complete matrix of interstitial macromolecules. The parenchyma accounts for a major part of the lungs' exposed surface. The injury response elicited by environmental toxins is a composite of the various cells present in the parenchyma. The simplest models for studying the effect of environmental toxins consist of isolated cells of various types, in culture. Because cell culture procedures permit the isolation and separation of different cell types, the effects of toxins on individual cell species in the lung can be delineated

under precisely controlled conditions. The use of cell cultures has provided fundamental information concerning biochemical mechanisms and
interrelationships between various intracellular metabolic processes and has
led to the elucidation of mechanisms of informational macromolecules and
protein synthesis. The applications of cell cultures in investigations on environmental toxicology have been reviewed.[8,20] Studies using cell cultures
also pointed out that the response of a tissue to an external stimulus is
not the sum of the individual reactions of its component cell types, but
is a concerted response resulting from complex interactions between the
different cell types and between cells and their surrounding extracellular
macromolecules.[21-24] The effect of an exogenous agent on a tissue can
be best examined under conditions in which all cells native to that tissue
are present in their natural apposition in relation to their neighboring cells
and matrix. For this reason, tissue explants (small fragments of tissue) can
be examined directly for their ability to interact with test environmental
agents. Explant procedures involve placing a piece of lung in a nutrient
medium under conditions where the cells remain viable for limited periods.
Because only very small amounts of tissue are needed, this procedure can
be adapted for biopsy specimens. Tissue explants reproduce the very earliest
response of the lungs in the intact animal and allow the examination of
the initial events in the interaction between the tissue and the toxin. The
limited viability of tissues under explant conditions precludes the examination of cellular events which span one or more cell cycles.

This is a disadvantage because many aspects of cell-environment interaction involve inductive processes that are not completed until the cell has
passed through the critical stages.[25,26]

LUNG ORGAN CULTURES

Many of the difficulties arising in cell culture or explant procedures are
avoided if the explants are maintained under conditions that permit them
to remain viable for longer periods, without losing the properties that
characterize the tissue. This is the basis of organ culture procedures. Organ
culture may be defined as a procedure in which a tissue can be maintained
under culture conditions for prolonged periods of time in its differentiated
state. By contrast to explant procedures, organ cultures permit the observation of the tissue for a longer period of time. Organ cultures usually do
not survive as long as isolated cells; however, in contrast to cell cultures,
which select out cell types, organ cultures maintain differentiated cells in
their native interrelations. Organ cultures not only maintain their histological
organization but grow in size and differentiate in a more or less normal
manner until the diffusion of nutrient medium and oxygen becomes limiting.

At this point, their integral biochemical processes begin to slow down and this may be considered to be an endpoint for the organ culture.

Organ culture procedures were first used nearly 50 years ago for the study of bone development.[27] Organ cultures have been used extensively for morphological[28-30] and physiological studies on vitamins[31,32] and hormones[33-35] for studying carcinogenesis[36] and embryonic induction.[37-39] Very few studies have utilized organ cultures of lungs despite the advantages offered by this system. One underlying reason for this may be that most of the studies on lung pathology have been concerned with lung function, but pulmonary physiology can hardly be examined in excised tissues. Lung organ cultures are not conducive to investigations of the mechanical aspects of lung function and are not subject to the rampant systemic physiology of the organism, having been cut off from the animal's circulation and immune system. Any biochemical effects of a test agent in a lung organ culture are part of the lungs' own inherent mechanisms to cope with injury and would provide the most direct evidence for the effect of the toxin on lungs. Our studies discussed in the following sections suggest that lung organ cultures are a useful model for investigating the biochemical aspects of pulmonary toxicology, and they may serve as a sensitive tool for evaluating the toxicity of known and suspected environmental toxins.

Lung Organ Culture: Method

Organ culture procedures are quite simple and do not require highly specialized equipment. The principal requirement for successful culture of differentiated tissues is that a compatible medium and sufficient oxygen be available and that culture conditions should resemble the tissues' normal physiological environment, as far as possible. The basic media for various types of organ cultures have been established on the basis of specific nutritional needs of different tissues. The availability of oxygen is particularly important since there is no circulation. In its simplest form, organ culture of lungs is carried out by placing small fragments of lung at the interphase between the nutrient medium and the gaseous phase. The size of the tissue fragments is important because it regulates the diffusion of the medium as well as of oxygen into the tissue; nutrients from the liquid medium and atmospheric oxygen must be freely available to all cells in the tissue. Placing the tissue at the interface is achieved by mounting it on a raft made of stainless steel grid or some other inert support.[40,41] In the procedure used in our laboratory,[42,43] lungs removed from newborn rats within two hours after birth are cut into strips (1 x 1 x 3 mm) and placed on Millipore filters which are gently floated over a 1-ml medium contained in the well of a disposable plastic (Falcon) organ culture dish. The filter paper liner of the surrounding chamber is saturated with sterile distilled water. The medium is the Dulbecco-Vogt modification of Eagle's minimum essential medium[44] with Earle's salt solution and contains, additionally, 90 μg/ml ascorbate and 2 mM glutamine. The medium also contains 10% fetal calf serum. We also routinely add penicillin, 50 U/ml and streptomycin, 50 μg/ml. The medium is replaced every other day unless otherwise warranted. A number of organ culture dishes are placed on shelves within a gas-tight plastic (Lucite) chamber. The bottom of the chamber is covered with a 5-mm layer of sterile water, to maintain a highly humid atmosphere. The inlet for the gas phase of the cultures is located at the bottom of the Lucite chamber. The incoming gas is moistened by being bubbled through the layer of water. Unless the composition of gaseous phase itself is being varied as part of the experiment, the gaseous phase consists of 5% CO_2 in air. The Lucite chamber is placed in a tissue culture incubator, maintained at 37°C. In addition to maintaining complete sterility at all steps, several precautions are necessary. The tissue slice must be thin enough to permit maximum permeability to the medium and gases. It is important that the tissue remain on top of the medium at all times and it should not be covered with medium; this limits the availability of oxygen and results in reduced viability.

MORPHOLOGY OF LUNG ORGAN CULTURES

The structure and growth of the lung in the intact animal are regulated by various physiological factors and mechanical constraints provided by the surrounding structures. Both lung structure and growth are altered in disease. The principal architectural feature of lungs related to their function is the presence of an extensively branched network of airways which culminate in thin-walled sacs—the alveoli. The rudiments of this organization are formed before birth but their final development is aided by the initiation of uninterrupted cyclic inflation and deflation at birth. None of these regulatory constraints are present in the organ culture system, and tissues maintained in this manner cannot be expected to show morphological characteristics of lungs observed *in vivo*. Since the major biochemical characteristics of lungs are a function of the differentiated states of their component cells, and are regulated by the interactions between different types of cells and between cells and their surrounding extracellular matrix, maintenance of these relationships in culture is important in studies on the biochemistry of toxin-tissue interaction in this model.

The normal rat lung is composed of more than 40 different cell types organized into complex tissue masses that form the structural support, vasculature and epithelial lining of these distal airways. Rapid changes take place near the end of gestation and immediately after birth. These changes are important in the maturation process. An understanding of the maturation process within the lung of newborn rats is required as a standard against which organ cultures can be compared and upon which an evaluation of the culture system can be made.

Figures 2-6 are light micrographs of neonatal rat lung and lung organ cultures. The tissues of lung culture prepared for microscopy could not be artificially inflated by injecting fluid fixative through the air passages at standard pressures. To facilitate comparision, the samples of control lung tissue removed from living rats have not been inflated. The ratio of air-passage volume to tissue volume changes very rapidly at the time of birth. Many of the alveoli in the lung are not fully inflated at the time of birth (Figure 2). When the lung of a rat is uniformly and fully inflated (a few hours after birth) the alveolar septa are very thin and contain little supportive tissue (Figure 3). Not all epithelial cells that line the air passages are fully differentiated at the time of birth (Figure 2), but the cuboidal and flattened alveolar cells can be distinguished in lungs of one-day-old rats (Figure 3). Hence, the lung tissue of newborn rats from which these cultures were made could be expected to change rapidly *in vivo*. Some changes within the tissue are similar in nature but are different in magnitude when the tissue is removed from the newborn rat and placed in culture.

All cell types in a tissue do not respond to culture conditions in the same way. Earlier studies on lung organ cultures indicated that while epithelial and connective tissue elements developed well, the growth of the vasculature was not comparable[45,46] and the alveolar spaces were obliterated by ingrowing cells.[40]

Our results show that as, in maturation *in vivo*, conditions of lung organ culture resulted in a similar proliferation of the connective tissues and the alveolar epithelium. A description of the structural changes in these two diverse tissue components may, therefore, serve as an abbreviated description of the response to culture conditions. The total amount of these tissues formed and rate of formation was much more rapid in the culture environment. The amount of connective tissue in the lung tissue after one-day culture (Figure 4) appeared to be greater than that found in the lung of three-day-old rats. Lung organ cultures were characterized by increasing numbers of epithelial cells, many of which were not fully differentiated and did not resemble the alveolar epithelial cells found on the alveolar surfaces in adult rats. These cells were associated with a highly distorted and thickened basal lamina. The vestiges of the alveolar spaces were reduced in size as the proliferating epithelial cells pushed inward and formed a more compact tissue mass. Small nests of these cells appeared to have large vacuolated structures in the cytoplasm (Figures 5 and 6). These cells increased in number with the duration of culture but the vesicles did not seem to have the same staining reaction as the normal secretory vesicles of the type II epithelial cells.

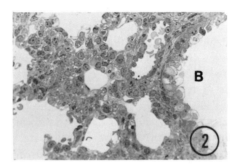

Figure 2. Alveoli of newborn rat lung do not inflate uniformly without special techniques of preparation. There is minimal connective tissue present in the distal airways including the wall of a tertiary bronchus (B). Few type II alveolar epithelial cells can be identified. (X 350)

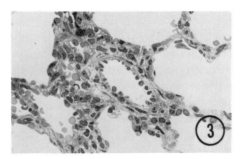

Figure 3. Lung of one-day-old rat shows uniformly inflated alveoli with thin alveolar septa. Many type I and type II alveolar epithelial cells can be distinguished (X 350).

We also compared the ultrastructures of lung organ cultures and lungs *in vivo*. Electron microscopic examination confirmed that the major differences between lung morphology *in vivo* and in organ cultures involved the epithelial and connective tissue structures. During *in vivo* lung growth, the epithelium lining the alveolar spaces is differentiated

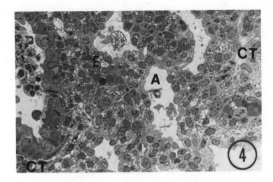

Figure 4. Vestiges of alveolar spaces (A) are retained after one day in culture. Some cells have many cytoplasmic vesicles (E). There is more connective tissue (CT) than found in the lung of one-day-old rat (X 350).

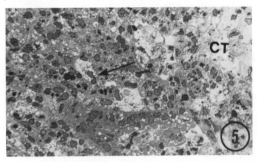

Figure 5. After two days in culture there are large amounts of connective tissue (CT) and few vestiges of alveolar spaces. The number of cells containing cytoplasmic vesicles (arrow) has increased (X 350).

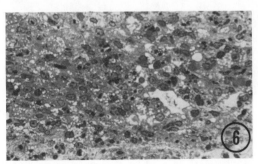

Figure 6. There are extensive areas of connective tissue which contain a few dispersed cells after three days in culture. Epithelial cells form compact tissue masses between the supportive tissues (X 350).

into two distinct cell types two days after birth. The flattened epithelial cells that cover most of the alveolar surface in the two-day-old rat are type I cells. They are characterized by few inclusions except pinocytotic vesicles and small mitochondria. The type II epithelial cells are cuboidal in shape, have short microvilli that extend from the apical surface into the alveolar space and contain many lamellar bodies in the cytoplasm. At the time of birth, the cells that line the alveolus are not always structurally distinguishable (Figure 7) and some cells with characteristics of type I cells may also contain lamellar bodies.

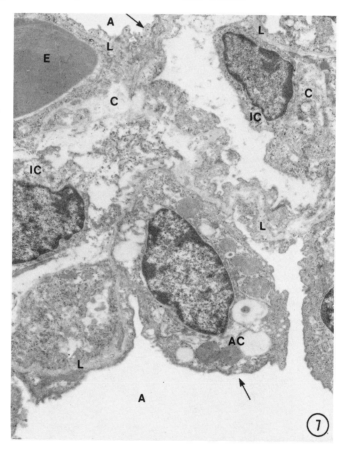

Figure 7. Electron micrograph of newborn rat lung. Alveolar septa (between the arrows) of a newborn rat are folded and contain small amounts of fibrous collagen (C). Interstitial cells (IC) have very irregular contours and are located between the basal laminae (L) of the opposing alveoli (A). The alveolar cell (AC) has very attenuated cytoplasmic processes that separate the alveolus from its underlying basal lamina. The erythrocyte (E) is located within a capillary of the septum (X 9,800).

The ultrastructure of supporting tissues of the lung also changes *in vivo* after birth. There is little evidence of fibrous collagen within the alveolar septa in the lungs of newborn rats (Figure 7) and larger numbers of collagen bundles may be seen in sections of lung from three-day-old rats (Figure 8).

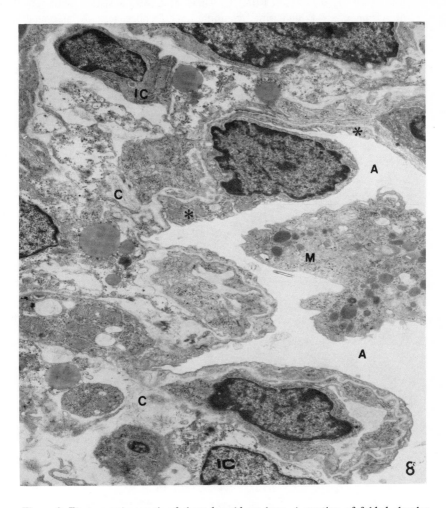

Figure 8. Electron micrograph of three-day-old rat lung. A portion of folded alveolar septum from a three-day-old rat contains both interstitial cells (IC) and small bundles of loosely organized collagen fibrils (C). The cell on the alveolar surface is a type I epithelial cell (*) with its characteristic attenuated cytoplasmic processes. A portion of a free alveolar macrophage (M) is shown within the alveolus (A) (X 7,400).

The alveolar epithelial cells and the interstitial components of the lung are also structurally changed in organ culture. Most of these structural changes may be attributed to a rapid proliferation of the same components that increase *in vivo*. As seen in Figure 9, there is much more fibrous collagen apparent in newborn rat lungs after three days in culture, than in lungs in three-day-old rats (Figure 8). These data are consistent with biochemical evidence for increasing accumulation of collagen in the neonatal lung *in vivo*.[47]

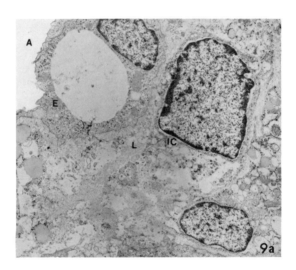

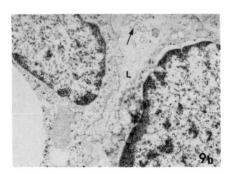

Figure 9. Electron micrograph of lung after three days in culture: (a) most alveolar epithelial cells (AC) are irregular in shape after three days in culture and set on a thickened basal lamina (X 7,400); (b) higher magnification (X 9,800). Arrow shows limited numbers of collagen fibers in the alveolar septum.

MACROMOLECULAR SYNTHESIS IN
LUNG ORGAN CULTURE

Morphological studies suggest that the fundamental interrelationships between various lung cells are retained in organ culture. This is reflected in the continued biochemical expression of their differentiated state, and, as discussed below, our cultures of neonatal lungs continued to synthesize the major macromolecules in a pattern paralleling macromolecular synthesis in lungs in newborn animals. Biochemical changes are a significant component of lung development and are involved in lung disease. As summarized in Figure 1, the interaction of a toxic agent with one or more tissue sites can cause major alterations in cellular metabolism, genetic function and macromolecular synthesis. Interference with the synthesis or degradation of major cellular and extracellular macromolecules, DNA, RNA and proteins including collagen, contributes significantly to the development of disease. It was of interest, therefore, to examine these biochemical events in lung cultures and to compare them with biochemical characteristics of healthy and diseased lungs.

The morphological development of lungs continues beyond birth in most mammalian species examined. In rats, enlargement of the lung and major structural changes in the gas-exchange region occur in the first four or five days, and new alveoli and septa continue to form between the fourth and thirteenth days.[48] These changes involve increases both in the numbers of cells and in the extracellular components. Although only morphological evidence is available concerning cell proliferation, it may be presumed that large amounts of DNA and protein accumulate during the early stages of neonatal lung development. During neonatal growth of the rabbit lung, the protein content on a dry weight basis does not change significantly whereas the collagen content undergoes a fivefold increase.[47] The rate of collagen synthesis per cell (DNA) is at a constant low level in the adult rabbit but is five times higher in the neonatal rabbit lung.[47] The higher rate of collagen synthesis in the early development of the rabbit lung also represents a marked emphasis on collagen synthesis in relation to total protein synthesis.

In our experiments, newborn rat lungs continued to "grow" and "develop" by these criteria, since they accumulated DNA, proteins and collagen, with an increasing emphasis on collagen synthesis per cell.[43,49]

Organ cultures of neonatal rat lungs showed sustained growth for more than five days. The rate of growth as measured by estimating the major macromolecular components was essentially linear over the five-day period and, in most of our experiments, we found this length of time to be suitable for studying biochemical processes as well as interactions of the cultures with toxic substances.[42,43]

Under our experimental conditions, cells in the cultured tissue continued to proliferate throughout the culture period as seen both by morphological and biochemical criteria.[43] As seen in Table III, DNA accumulated in the cultures and increased nearly 50% at the end of five days. The total protein content kept up with the DNA content, and increased to the same extent during this period.

Table III. DNA and Protein Contents in Organ Cultures of Lung

| Macromolecules | Hours in Cultures | | |
	24	72	120
DNA (μg/mg dry wt.)	33.2	44.0	47.8
Protein (μmoles/mg dry wt.)	690	871	1043

Each number represents the average of six cultures.

Protein is expressed as μmoles of leucine equivalent in the acid hydrolyzate of the tissue by following the amount of total ninhydrin reactive material using leucine as standard.

Lung cultures were grown as described in the Methods Section. At each indicated time, tissue samples were analyzed for DNA and total protein content. Dry weight was determined after lyophilizing the tissue samples to a constant weight.

These cultures also synthesized and accumulated collagen as determined by hydroxyproline assays (Table IV). The hydroxyproline content increased both in relation to the dry weight of the tissue and in relation to the total protein content, indicating that there was a specific increase in the proportion of collagen during this period of growth. Since the numbers of cells in the cultured tissues increased continuously as seen by the increase in the DNA content, it was of interest to determine if the collagen output per cell remained constant or increased. As seen in Table IV, the hydroxyproline content per unit DNA was increased. In these experiments no attempt was made to separate collagen from noncollagenous proteins, because of the difficulties encountered in extracting lung collagens.[8] Since hydroxyproline occurs only in collagen, a ratio of hydroxyproline content to the total protein content is a measure of the fraction of total protein accounted for by collagen. As seen in Table IV, this ratio increased consistently throughout the culture period indicating that more collagen was accumulating in the culture than were noncollagenous proteins. These data, along with the data on the cellular output of collagen, indicate that cells in the organ cultures continued to differentiate into new connective tissue and are consistent with observations on neonatal lung growth *in vivo*.

Table IV. Hydroxyproline Content in Organ Cultures of Lung

Hours in Culture	Hydroxyproline Content	
	μg/mg dry wt.	μg/leu. equiv.
24	1.88	0.42
72	2.75	0.55
120	3.15	0.63

Each number represents the average of six cultures.

Leucine equivalent was determined in the acid hydrolyzate of the tissue by following the amount of total ninhydrin reactive material using leucine as standard.

Hydroxyproline was estimated by the standard colorimetric method after hydrolyzing tissue samples with $6 N$ HCl at $120°C$ for 20 hr. Dry weight was taken as mentioned in Table III.

Collagen represents a family of tissue-specific proteins.[50] Because lungs contain all types of tissue organization, they have been shown to contain all different known collagen chains.[8] The major collagen component of the lung parenchyma is type I collagen, accounting for nearly 70% of the total, and type III collagen, accounting for nearly 30%, and these proportions change in disease.[8,51] We examined the nature of collagen synthesized in the lung organ cultures.[52] After three days in cultures, the collagen synthesized by newborn rat lungs consisted of 69% type I and 31% type III, strongly suggesting that lung cells remain in their normal differentiated state in organ culture.

RESPONSE OF ORGAN CULTURES TO CHEMICAL INJURY

The earliest events in the initiation of pulmonary injury by toxic agents are biochemical. Because they retain normal biochemical characteristics, lung organ cultures are particularly suited for the examination of the biochemical effects of toxic agents and for determining the molecular mechanisms involved in these interactions. To evaluate the applicability of this system in investigations of environmental lung injury, we examined the biochemical effects of several known pulmonary toxins. The biochemical response of lung organ cultures to several unrelated pulmonary toxins mimicked the response of lungs in vivo to these agents. Some of our studies are discussed below.

Effect of High Oxygen Concentrations

Exposure of lungs to oxidant gases such as ozone (O_3) and nitrogen dioxide (NO_2) and to high concentrations of oxygen (O_2), leads to major biochemical and morphological alterations.[4,6,53-56] The squamous epithelial lining of the alveoli is damaged and replaced by increasing number of thick cuboidal cells, and large amounts of fibrous collagen are deposited in the alveolar interstitium.

We examined the effect of 95% O_2 environments on lung organ cultures. Cultures maintained in 95% O_2 were characterized by a proliferation of type II cells, which lined the alveoli in a single layer. These cuboidal cells had short microvilli on their apical surfaces and contained large numbers of lamellar lipid bodies in their cytoplasm (Figure 10a). These cultures were also characterized by large deposits of fibrous collagen seen clearly in the higher magnification electron micrograph (Figure 10b).

When lung organ cultures were maintained in 95% O_2, they exhibited decreased cellularity. As seen in Table V, the DNA content of control cultures continued to increase throughout the five-day culture period. By contrast, in cultures maintained in 95% O_2, there was an initial increase both in the DNA and protein,[43] reflecting increased cellular proliferation (see Table VIII). After 48 hours, however, there was a marked decrease in both parameters, although the ratio of protein to DNA remained high. These data are consistent with the observation that DNA synthesis is markedly decreased in lungs of animals exposed to high O_2.[56]

The total protein (noncollagen + collagen) content per mg dry weight of the cultures increased in the controls throughout the five-day culture period; however, there was only an insignificant increase in protein content on this basis in cultures maintained in 95% O_2 (Table VI). Comparison of the control and oxygen treated cultures indicated that there was much less accumulation of protein in the latter, being 85% of the control after three days and 70% after five days in culture. We also related the total protein content to the DNA content, that is the cellularity of the cultures. As seen in Table VII, there was an increase in the total protein content on this basis at all times examined. This could be attributed to the fact that in the control cultures, both the DNA and the total protein contents increased at nearly the same rate so that the ratio of protein to DNA remained nearly constant. By contrast, in the cultures exposed to O_2, the DNA content declined sharply (Table V), whereas as seen in Table VI, the protein content was not changed significantly. Thus, there was a greater apparent accumulation of protein in the oxygen-exposed cultures on a per cell basis even though the net protein content did not change appreciably.

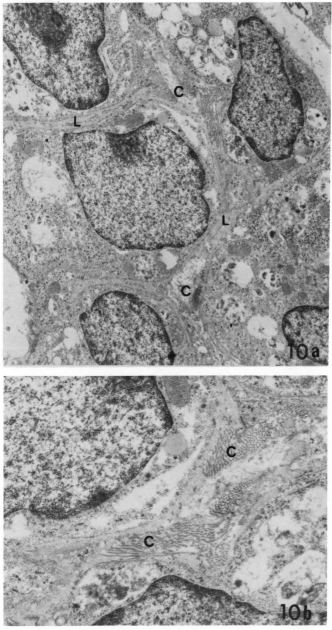

Figure 10. (a) Electron micrograph of lung cultures maintained in 95% oxygen for three days. Basal lamina (L) is much thicker than in 3-day control cultures. Collagen bundles (C) are apparent (X 7,400); (b) higher magnification (X 9,800) of a portion of Figure 10a showing well-formed collagen bundles.

Table V. DNA Content of Lung Organ Cultures in the Presence of 95% Oxygen

Hours in Culture	DNA Content (μg/mg dry wt.)	
	Control	95% O_2
24	36.75	64.0
72	41.50	40.4
120	49.13	31.8

Each number represents the average of six cultures.
Lung cultures were grown in the presence of 95% O_2+5% CO_2 and in 95% air +5% CO_2 (control). At each indicated time, samples were analyzed and dry weight taken as described in Table III.

Table VI. Protein Content of Lung Organ Cultures
Grown in the Presence of 95% Oxygen

Hours in Culture	Protein Content (μmole/leu. equiv./mg dry wt.)	
	Control	95% O_2
24	718	702
72	855	718
120	1043	735

Each number represents the average of six cultures.
Lung cultures were grown and analyzed for total protein as indicated in Tables III and V.

Table VII. Protein and Hydroxyproline Contents per Cell
in Organ Cultures of Lung Grown in the Presence of 95% Oxygen

Hours in Culture[b]	Macromolecule Content	
	Protein/DNA	Hydroxyproline/DNA[d]
Control:		
24	6	1.13
72	6.5	1.35
120	7.3	1.43
95% O_2		
24	10.5	1.49
72	14.2	2.62
120	24.8	3.74

[a]Each number represents the average of five cultures.
[b]Methods for culturing lungs and analyses were as indicated in Tables IV-VI.
[c]Protein is expressed as μmoles leucine equivalent/DNA/mg dry wt.
[d]Hydroxyproline is expressed as μg/DNA/mg dry wt.

The collagen content per mg dry weight of control cultures increased significantly in the control cultures, and, after five days in culture, was 70% greater than after one day in culture. By contrast, in cultures exposed to 95% oxygen, the net collagen content per mg dry weight of the lungs did not increase during this period. As seen in Table VII, both total protein and the collagen content per mg DNA appeared to be increased per cell. The morphological correlate of this observation is the greater area occupied by interstitial collagenous material. In intact animal experiments these morphological observations are interpreted as "fibrosis." The observation that the collagen concentration as a fraction of total protein present in the tissue does not change markedly in morphologically "fibrotic" tissues, is consistent with the *in vivo* findings in idiopathic human pulmonary fibrosis[8] and in experimentally induced pulmonary fibrotic lesions in dogs.[14]

MACROMOLECULAR SYNTHESIS IN LUNG ORGAN CULTURES IN THE PRESENCE OF TOXIC METALS

Chronic exposure to vapors containing mercury or cadmium, or their compounds, results in pulmonary lesions characterized by fibrosis.[57,58] The metals are rapidly oxidized in the body and their toxic effects have been attributed to their ions. These metals interact with a large number of enzymes and proteins, forming tight complexes with -SH groups,[59,60] and destabilize lysosomal membranes, resulting in the release of lysosomal hydrolases.[61] They have also been shown to impair ion transport through membranes.[62] Cadmium ion (Cd^{2+}) decreases glucose metabolism[63] and adversely affects the respiration of pulmonary macrophages by uncoupling oxidative phosphorylation and inhibiting mitochondrial oxygen uptake.[64]

We examined the effects of these metals on macromolecular synthesis in lung organ cultures.[42,65] Concentrations of these ions as low as 0.1 μM markedly increased the synthesis of DNA, RNA and collagen.

When organ cultures were maintained in medium containing 0.1-10.0 μM Hg^{2+} (added as $HgCl_2$), they appeared to synthesize DNA (Figure 11) and RNA (Figure 12) at elevated rates after one day of exposure to the toxic ion. The rate of DNA synthesis, measured as the specific activity of DNA pulse-labeled with ^{3}H-thymidine, was elevated 2.5-3.5 times after 24 hours of exposure, but decreased over the five-day culture period until the cultures maintained at 1.0 or 10.0 μM Hg^{2+} showed less incorporation than the controls (Figure 11). These data point to the initiation of an early injury and reparative response in the cultures. The concentrations of Hg^{2+} used in these cultures were apparently cytotoxic as seen by the decreasing specific activity of DNA synthesis. A similar pattern of initial increase followed by a decrease was observed in the rate of synthesis of RNA, seen in the specific activity of RNA pulse-labeled in the presence of ^{3}H-uridine (Figure 12).

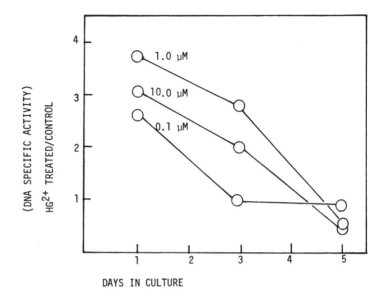

Figure 11. DNA synthesis in lung cultures exposed to Hg^{2+}. Lung cultures were grown in the presence of indicated concentrations of $HgCl_2$ added in the culture medium. At each indicated time, cultures were labeled with 10 μCi of ^{3}H-thymidine (10.6 Ci/mmol) for six hours. Tissue samples were analyzed for DNA using standard procedures after removing free radioactivity. Each point represents an average of three cultures.

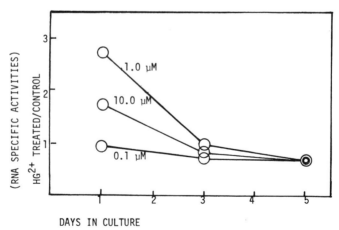

Figure 12. RNA synthesis in lung cultures exposed to Hg^{2+}. Lung cultures were grown in the presence of indicated concentrations of $HgCl_2$ added in the culture medium. At each indicated time, cultures were labeled with 10 μCi of 3H-uridine (27 Ci/mmol) for six hours. Samples were analyzed for RNA using standard procedure after removing free radioactivity. Each point represents an average of three cultures.

The synthesis of collagen was examined by following the incorporation of ^{14}C-proline into ^{14}C-hydroxyproline, after pulse-labeling at different times after continuous exposure to Hg^{2+}. The ratios of the specific activities of ^{14}C-hydroxyproline synthesized in the presence of various concentrations of Hg^{2+}, to the specific activity in controls, are presented in Table VIII. At each time point, the Hg^{2+}-treated cultures incorporated greater amounts of ^{14}C-proline into hydroxyproline, indicating enhanced rates of collagen synthesis. The greatest increase was observed at the lowest Hg^{2+} concentration used (0.1 μM), and in these cultures the rate of collagen synthesis continued to increase during the five-day culture period. At higher concentrations, however, the rate of collagen synthesis declined markedly on the fifth day, consistent with the cytotoxicity discussed above.

Table VIII. Synthesis of ^{14}C-Hydroxyproline in
Lung Organ Cultures Grown in the Presence of Hg^{2+}

Hours in Culture	Relative Specific Activity		
	0.1 μM	1.0 μM	10.0 μM
24	1.33	1.11	1.23
72	2.03	1.73	1.64
120	2.73	1.13	1.24

Each number represents the average of four cultures.
Ratio of DPM in ^{14}C-Hydroxyproline in treated vs control cultures.
Lung cultures were grown in the presence of indicated concentrations of $HgCl_2$ in the medium. At each indicated time, cultures were labeled with 5 μCi of (U)-^{14}C-proline (254 mCi/mmol) for 6 hr. Tissue samples were homogenized and dialyzed against water and hydrolyzed as mentioned in Table IV.
[14]C-hydroxyproline was estimated in the acid hydrolyzates by the standard procedure described previously.[12,29]

As seen in Table IX, the largest increase in the synthesis of collagen in cultures treated with Cd^{2+} was seen after 24 hours of exposure. At the lowest concentration (0.1 μM) used, the rate of collagen synthesis decreased after the first 24 hours, but remained higher than in the controls. At 1.0 or 10.0 μM Cd^{2+}, there was a marked decrease after the initial stimulation, and after 72 hours exposure, the exposed cultures synthesized less collagen than did the control cultures, indicating that this metal may be more toxic than mercury.

Table IX. Synthesis of ^{14}C-Hydroxyproline in
Lung Organ Cultures Grown in the Presence of Cd^{2+}

Hours in Culture	Relative Specific Activity[a]		
	0.1 μM	1.0 μM	10.0 μM
24	3.12	2.50	1.93
72	2.00	0.75	0.73
120	1.18	0.70	0.67

[a]Ratio of DPM in ^{14}C-hydroxyproline in treated vs control cultures.
Each number is an average of four cultures.
Lung cultures were grown in the presence of the indicated concentrations of $CdCl_2$ in the medium. Procedures for labeling and analysis of ^{14}C-hydroxyproline were as mentioned in Table VIII.

DISCUSSION AND CONCLUSIONS

The growing complexity of atmospheric pollutants with the continuous influx of new chemical toxins makes it imperative that rapid and reproducible experimental procedures be developed to assess their toxicity. As discussed in earlier sections, a variety of experimental approaches have been utilized for this purpose. The physiological complexity of intact animals, coupled with their inherent biological variability and their increasing susceptibility to infectious disease and nutrition-related disorders as a result of experimental procedures, make them a relatively inefficient system for routine testing procedures. Toxicity testing with animal models is also time-consuming, expensive and complicated by the fact that adverse biological effects of a test substance may require long periods to develop, and because toxicity tends to be cumulative, even when different insults are administered.[66] The delayed development of toxic effects also results in the superimposition of age-related physiological and functional morphological alterations in many experimental animals.

Organ cultures present a highly accessible *in vitro* model which avoids the shortcomings of intact animal studies. Organ cultures can be maintained in highly controlled environments and are less susceptible to infectious and nutritional disturbances than intact animals. Because of their isolation from the body, they are not subject to homeostatic mechanisms, circulating cells and proteins arising in distant parts of the body, hormonal influences or immune phenomena. Thus, a direct observation of the effect of an added exogenous agent represents the direct response of the cultured tissue to this insult.

Our studies show that lung organ cultures mimic the response of lungs *in vivo* to known pulmonary toxins. These responses are expressed as morphological alterations paralleling alterations seen in lungs exposed *in vivo* to the same toxins. They are seen much more clearly as biochemical alterations, involving macromolecular synthesis, and in collagen content.[42,43,49] These changes are a result of genetic-level alterations in the cells in the tissue explant in culture, and are also seen as alterations in the content of specific tRNA accompanying the collagen changes.[67] These characteristic events in lung injury and repair are observed in a very short time span in organ culture. Lung organ cultures are, therefore, a very good model for evaluating the toxicity and the mechanism of injury of those known and unknown agents that can be added to the culture environment.

ACKNOWLEDGMENTS

We wish to acknowledge the expert technical assistance provided by Maximita Tolentino, Belma Enriquez, Lynne Calonico and Nancy Crise-Benson. We thank Sandra Hodess and Shirley Rappaport for their assistance in the preparation of the manuscript. These studies greatly benefited from discussions with Dr. M. G. Mustafa and Dr. S. D. Lee. Support for this research was provided by grant HL-19668 from the National Heart, Lung and Blood Institute, by Contract 68-03-2005 from the U.S. Environmental Protection Agency and by grants from the California Lung Association.

REFERENCES

1. Staub, N. C. In: *Environmental Factors in Respiratory Disease*, D. H. K. Lee, Ed. (New York: Academic Press, 1972), p. 16.
2. Witschi, H. In: *Essays in Toxicology*, Vol. 6, W. J. Hayes, Ed. (New York: Academic Press, 1975), p. 125.
3. Witchi, H. "Environmental Agents Altering Lung Biochemistry," *Fed. Proc.* 36:1631 (1975).
4. Mustafa, M. G., A. D. Hacker, J. J. Ospital, M. Z. Hussain and S. D. Lee. In: *Biochemical Effects of Environmental Pollutants*, S. D. Lee, Ed. (Ann Arbor, MI: Ann Arbor Science Publishers, Inc., 1977), p. 59.
5. Mudd, J. B., and B. A. Freeman. In: *Biochemical Effects of Environmental Pollutants*, S. D. Lee, Ed. (Ann Arbor, MI: Ann Arbor Science Publishers, Inc., 1977), p. 97.
6. Chvapil, M. "Pharmacology of Fibrosis: Definitions, Limits and Perspectives," *Life Sci.* 16:1345 (1975).
7. Cross, C. E., A. K. Reddy, M. Z. Hussain, L. W. Schwartz, M. G. Mustafa, C. K. Chow and D. L. Dungrowth. "Long Adaptation to

Low Levels of Ozone on Long Term Exposure" *Am. Rev. Resp. Dis.* 113(4):102 (1972).

8. Hance, A. J., and R. G. Crystal. "The Connective Tissue of the Lung," *Am. Rev. Reps. Dis.* 112:657 (1975).

9. Chvapil, M. "Pharmacology of Fibrosis: Tissue Injury," *Environ. Health Pers.* 9:283 (1974).

10. Scheel, L. D., O. J. Dobrogorski, J. J. Mountain, J. L. Svirbely, and H. E. Stokinger. "Physiologic, Biochemical, Immunologic and Pathologic Changes Following Ozone Exposure," *J. Appl. Physiol.* 14:67 (1959).

11. Pariente, R. "Action of Oxygen and Oxidizer on the Lung and Bronchioles," *Prog. Resp. Res.* 8:91 (1975).

12. Hussain, M. Z., C. E. Cross, M. G. Mustafa, and R. S. Bhatnagar. "Hydroxyproline Content and Prolyl Hydroxylase Activities in Lungs of Rats Exposed to Low Levels of Ozone," *Life Sci.* 18:897 (1976).

13. Chvapil, M., and Y. M. Peng. "Oxygen and Lung Fibrosis," *Arch. Environ. Health* 30:528 (1975).

14. Orthoefer, J. G., R. S. Bhatnagar, A. Rahman, Y. Y. Yang, S. D. Lee and J. F. Stara. "Collagen and Prolyl Hydrolxylase Levels in Lungs of Beagles Exposed to Air Pollutants," *Environ. Res.* 12:299 (1976).

15. Fulmer, J. D., and R. G. Crystal. In: *The Biochemical Basis of Pulmonary Function,* R. G. Crystal, Ed. (New York: Marcel Dekker, Inc., 1976), p. 419.

16. Fulmer, J. D., R. S. Bienkowski, M. J. Cowan, K. H. Bradley, W. C. Roberts and R. G. Crystal. "Comparsion of Collagen Concentration, Distribution and Synthesis in Fibrotic and Normal Lungs," *Clin. Res.* 24:384A (1976).

17. Hance, A. J., A. L. Horwitz, M. J. Cowan, N. A. Elson, J. F. Collins, R. S. Bienkowski, K. H. Bradley, S. McConnel-Breul, W. M. Wagner and R. G. Crystal. "Biochemical Approaches to the Investigation of Fibrotic Lung Disease," *Chest* 69:(Suppl.) 257 (1976).

18. Wusteman, F. S., D. B. Johnson, K. S. Dodgson, and D. P. Bell. "The Use of Normal Rats in Studies on the Acid Mucopolysaccharides of Lung," *Life Sci.* 7:1281 (1968).

19. Adamson, I. Y. R., D. H. Bowden, and J. P. Wyatt. "A Pathway to Pulmonary Fibrosis: An Ultrastructural Study of Mouse and Rat Following Radiation to the Whole Body and Hemithorax," *Am. J. Pathol.* 58:481 (1970).

20. Collins, J. F. and R. G. Crystal. In: *The Biochemical Basis of Pulmonary Function,* R. G. Crystal, Ed. (New York: Marcel Dekker, Inc., 1976), p. 171.

21. Grobstein, C. In: *Extracellular Matrix Influences of Gene Expression,* H. C. Slavkin and R. C. Grenlich, Eds. (New York: Academic Press, 1975).

22. Moscona, A. A. In: *Extracellular Matrix Influences of Gene Expression,* H. C. Slavkin and R. C. Grenlich, Eds. (New York: Academic Press, 1975), p. 57.

23. Bernfield, M. R., and J. J. Cassiman. In: *Extracellular Matrix Influences of Gene Expression,* H. C. Slavkin and R. C. Grenlich, Eds. (New York: Academic Press, 1975), p. 457.

24. Reddi, A. H. In: *Biochemistry of Collagen*, G. N. Ramachandran and A. H. Reddi, Eds. (New York: Plenum Press, 1976), p. 449.

25. Steinberg, R. A., W. A. Scott, B. B. Levinson, R. D. Ivarie and E. M. Tomkins. In: *Regulation of Gene Expression in Eukaryotic Cells*, M. Harns and B. Thomson, Eds. DHEW Publ. #NIH 74-648 (1973), p. 53.

26. Holzer, H., S. Dienstein, J. Biehl, and S. Holtzer. In: *Extracellular Matrix Influences on Gene Expression*, H. C. Slavkin and R. C. Grenlich, Eds. (New York: Academic Press, 1976), p. 253.

27. Strangeways, T. S. P., and H. B. Fell. "Experimental Studies on the Differentiation of Embryonic Tissues Growing *In Vivo and In Vitro*," *Proc. Roy. Soc. London (B)* 99:340 (1926).

28. Saxen, L. "Effect of Tetracycline on Osteogenesis *In Vitro*," *J. Exp. Zool.* 162:269 (1966).

29. Bhatnagar, R. S., K. I. Kivirikko, and D. J. Prockop. "Studies on the Synthesis and Intracellular Accumulation of Protocollagen in Organ Culture of Embryonic Cartilage," *Biochim. Biophys. Acta* 154:196 (1968).

30. Kaitila, I., J. Wartiovaava, O. Laitinen, and L. Saxen. "The Inhibitory Effect of Tetracycline on Osteogenesis in Organ Culture," *J. Embryol. Exp. Morphol.* 23:185 (1970).

31. Kochhar, D. M., J. T. Dingle, and J. A. Lucy. "The Effects of Vitamin A on Cell Growth and Incorporation of Labelled Glucosamine and Proline by Mouse Fibroblasts in Culture," *Exp. Cell Res.* 52: 591 (1968).

32. Jeffrey, J. J., and G. R. Martin. "The Role of Ascorbic Acid in the Biosynthesis of Collagen," *Biochim. Biophys. Acta* 121:269 (1966).

33. Alescio, T., and A. M. Dani. "The Influences of Mesenchyme on the Epithelial Glycogen and Budding Activity in Mouse Embryonic Lung Developing *in Vitro*," *J. Embryol. Exp. Morphol.* 25:131 (1971).

34. Stern, P. H., and L. G. Raisz. "An Analysis of the Role of Serum in Parathyroid Hormone-Induced Bone Resorption in Tissue Culture," *Exp. Cell. Res.* 46:106 (1967).

35. Adamson, I. T. R., and D. H. Bowden. "Reaction of Cultured Adult and Fetal Lung to Prednisolone and Thyroxine," *Arch. Pathol.* 99: 80 (1975).

36. Diamond, L., and W. M. Baird. In: *Growth, Nutrition and Metabolism of Cells in Culture*, G. H. Rothblat and V. Cristofalo, Eds. (New York: Academic Press, 1977), p. 421.

37. Grobstein, C. "Mechanisms or Organogenetic Tissue Interaction," *Nat. Cancer Inst. Monogr.* 26:279 (1967).

38. Croissant, R., H. Guenther, and H. C. Slavkin. In: *Extracellular Matrix Influence on Gene Expression*, H. C. Slavkin and R. C. Greulich, Eds. (New York: Academic Press, 1975), p. 515.

39. Rutter, W. J., R. L. Pictet, and P. W. Morris. "Toward Molecular Mechanisms of Developmental Processes," *Ann. Rev. Biochem.* 42: 601 (1973).

40. Trowell, O. A. "The Structure of Mature Organs in a Synthetic Medium," *Exp. Cell Res.* 16:118 (1959).

41. Adamson, I. Y. R., and D. H. Bowden. "The Intracellular Site of Surfactant Synthesis: Autoradiographic Studies on Murine and Avian Lung Explants," *Exp. Mol. Pathol.* 18:112 (1973).

42. Hussain, M. Z., R. S. Bhatnagar and S. D. Lee. In: *Biochemical Effects of Environmental Pollutants,* S. D. Lee, Ed. (Ann Arbor, MI: Ann Arbor Science Publishers, Inc., 1977), p. 341.

43. Hussain, M. Z., J. C. Belton, and R. S. Bhatnagar. "Macromolecular Synthesis in Organ Cultures of Neonatal Rat Lung," *In Vitro,* 14:740 (1978).

44. Smith, J. D., G. Freeman, M. Vogt, and R. Dulbecco. "The Nucleic Acid of Polyoma Virus," *Virology* 12:185 (1960).

45. Sorokin, S. P. "A Study of the Development in Organ Cultures of Mammalian Lungs," *Develop. Biol.* 3:60 (1961).

46. Rosin, A. "Studies on Lung Tissue *In Vitro* With Special Reference to the Nature of the Lining Cells of the Alveolus," *Anat. Rec.* 97: 447 (1947).

47. Bradley, K. H., S. D. McConnell, and R. G. Crystal. "Lung Collagen Composition and Synthesis," *J. Biol. Chem.* 249:2674 (1974).

48. Burri, P. H., and E. R. Weibel. In: *Development of the Lung,* W. A. Hodson, Ed. (New York: Marcel Dekker, 1977), p. 215.

49. Hussain, M. Z., J. C. Belton, and R. S. Bhatnagar. In: *Proc. 4th Joint Conference on Sensing of Environmental Pollutants, Am. Chem. Soc.* 532-536 (1978).

50. Miller, E. J. "Biochemical Characteristics and Biological Significance of the Genetically Distinct Collagens," *Mol. Cell. Biochem.* 13:165 (1976).

51. Sayer, J. M., E. T. Hutcheson, and A. H. Kang. "Collagen Polymorphism in Idiopathic Chronic Pulmonary Fibrosis," *J. Clin. Invest.* 57: 1498 (1976).

52. Bhatnagar, R. S., M. Z. Hussain, J. A. Streifel, M. Tolentino and B. Enriquez. "Alteration of Collagen Synthesis in Lung Organ Cultures by Hyperoxic Environments," *Biochem. Biophys. Res. Commun.* 83:392 (1978).

53. Adamson, I. Y. R., and D. H. Bowden. "Pulmonary Injury and Repair," *Arch. Pathol. Lab. Med.* 100:640 (1976).

54. Stephens, R. J., M. F. Sloan, M. J. Evans, and G. Freeman. "Early Response of Lung to Low Level of Ozone," *Am. J. Pathol.* 74: 31 (1974).

55. Adamson, I. Y. R., D. H. Bowden, and J. P. Wyatt. "Oxygen Poisoning in Mice: Ultrastructural and Surfactant Studies During Exposure and Recovery," *Arch. Pathol.* 92:279 (1971).

56. Bowden, D. H., and I. Y. R. Adamson. "Reparative Changes Following Pulmonary Cell Injury: Ultrastructural, Cytodynamic and Surfactant Studies in Mice After Oxygen Exposure," *Arch. Pathol.* 92: 279 (1971).

57. Tennant, R., H. J. Johnson, and J. B. Wells. "Acute Bilateral Pneumonitis Associated with the Inhalation of Mercury Vapors," *Conn. Med.* 25:106 (1961).

58. Hayes, J. A., G. L. Snider and K. C. Palmer. "The Evolution of Bio-Chemical Damage in the Rat Lung After Acute Cadmium Exposure," *Am. Rev. Resp. Dis.* 113:121 (1976).

59. Brenner, I. "Heavy Metal Toxicities," *Quart. Rev. Biophys.* 7:75 (1974).
60. Hughes, W. L., Jr. "Protein Mercaptides," *Cold Spring Harbor Symp. Quant. Biol.* 14:79 (1950).
61. Taylor, N. S. "Historical Studies of Nephrotoxicity With Sublethal Doses of Mercury in Rats," *Am. J. Pathol.* 46:1 (1965).
62. Fleisher, L. N., T. Yorio, and P. J. Bentley. "Effect of Cadmium on Epithelial Membrane," *Toxicol. Appl. Pharmacol.* 33:384 (1975).
63. Ithakossis, D. S., T. Ghafgazi, J. A. Alennear, and W. V. Kessler. "Effect of Multiple Doses of Cadmium on Glucose Metabolism and Insulin Secretion in the Rat," *Toxicol. Appl. Pharmacol.* 31:143 (1975).
64. Mustafa, M. G., and C. E. Cross. "Pulmonary Alveolar Macrophage: Oxidative Metabolism of Isolated Cells and Mitochondria and Effects of Cadmium Ion on Electron and Energy Transfer Reactions," *Biochemistry* 10:4176 (1971).
65. Hussain, M. Z., and R. S. Bhatnagar. "Macromolecular Synthesis in Lung Organ Cultures Treated with Mercuric Ions," In press.
66. Condouris, G. A. In: *Drill's Pharmacology in Medicine, 4th ed.* J. R. DiPalma, Ed. (New York: McGraw Hill Book Co., 1971), p. 20.
67. Hussain, M. Z., B. Enriquez, M. Tolentino, and R. S. Bhatnagar. "Stimulation of Collagen-Related tRNA Levels in Lung Organ Cultures," *Fed. Proc.* 36:607 (1977).

CHAPTER 9

REACTION OF OZONE WITH HUMAN ERYTHROCYTES

B. A. Freeman, B. E. Miller and J. B. Mudd
Department of Biochemistry
Statewide Air Pollution Research Center
University of California
Riverside, California 92502

ABBREVIATIONS

AChE, acetylcholinesterase; Chol, cholesterol; GPDH, glyceraldehyde-3-phosphate dehydrogenase; G6PD, glucose-6-phosphate dehydrogenase; IO vesicles, inside out vesicles; PC, phosphatidylcholine; PE, phosphatidylethanolamine; ppm, parts per million (or $\mu l/l$); PPS, pentose phosphate shunt; PS, phosphatidylserine; RBC, red blood cells; RO vesicles, right side out vesicles; SM, sphingomyelin; TBA, thiobarbituric acid; TCA, trichloracetic acid; TLC, thin layer chromatography.

INTRODUCTION

Ozone is the major oxidant of photochemical air pollution. Ground levels of ozone range from a background of 0.02-0.70 ppm in polluted air.[1] Ambient concentrations of ozone affect primarily the tracheobronchial regions of exposed primates and rodents, causing alterations in both the morphology and dynamic function of the respiratory system. Ozone has also been reported to be responsible for various extrapulmonary effects. These include changes in the specific activity of certain erythrocyte enzymes,[2,3] and increases in the phenobarbital sleeping time of ozone-exposed mice, suggesting an interaction with liver cytochrome P-450 oxidase function.[4]

Several authors have concluded that cytotoxicity of ozone is due primarily to a reaction with membrane lipids.[2,5-9] On the other hand,

151

experiments in our laboratory employing the unilamellar[10] and multi-lamellar[11] species of egg phosphatidylcholine bilayers, and protein-containing biomembranes[12,13] have shown that olefinic groups located in the hydrophobic regions of membranes are resistant to ozonolysis. We wish to resolve these differences. The first experiments examine the influence of ozone on membrane integrity and composition. The capability of ozone to cross cell membranes is also tested.

The second group of experiments deals with the interaction between ozone and the enzymes of the pentose phosphate shunt (PPS), which are responsible for maintaining adequate levels of cytoplasmic-reduced glutathione (GSH). Other investigators have reported increases in the specific activities of glucose-6-phosphate dehydrogenase, glutathione peroxidase and glutathione reductase in lung homogenates of rats or monkeys breathing low levels of ozone.[14-17] It is not clear whether the increases in these enzyme activities are due to an induction of enzyme synthesis, the infiltration of inflammatory cells into damaged lung tissue[18] or the proliferation of alveolar type II cells seen to occur subsequent to ozone damage of alveolar type I cells. We wished to examine the effects of ozone on this enzyme system in a single-cell type, such as the erythrocyte, which would allow elucidation of the influence which ozone has on cellular metabolism without the complications of enzyme induction and damage of cell type. These experiments were also intended to aid interpretation of the report of Buckley *et al.,* who observed elevated erythrocyte glucose-6-phosphate dehydrogenase levels and a decreased GSH content in blood drawn from human volunteers breathing 0.5 ppm ozone for 2 3/4 hours.[3]

EXPERIMENTAL PROCEDURES

Ozone Exposure

Ozone was generated in an oxygen stream flowing at 20 ml/min through an apparatus producing a silent electric discharge or containing a mercury lamp. Ozone concentration in the gas stream was monitored by bubbling the gas through a 5.0-ml neutral buffered KI solution, which was 0.06 M with respect to pH 7.0 phosphate buffer and 0.04 M with respect to KI.[19] The liberated iodine was assayed spectrophotometrically at 350 nm using I_2 E_M = 24.2 x 10^3.[20]

Lipid Analysis

Lipids were chromatographed on 0.3-mm-thick thin layers of Silica Gel G (E. M. Laboratories, Elmsford, NY). Two-dimensional development of lipids used chlorofrom-methanol-7 N ammonium hydroxide (65:30:4) in the first dimension and chloroform-methanol-acetic acid-water (170:25: 25:6) in the second dimension.[21] Plates were sprayed with 0.01% 8-anilino-1-naphthalene sulfonic acid and lipids detected by quenching of fluorescence under UV light. Phosphorus in lipid fractions was determined by the method of Ames[22] after sample digestion with ethanolic $Mg(NO_3)_2$. Cholesterol was determined by the o-phthalaldehyde method.[23] Fatty acid methyl esters were made by the method of Beare-Rodgers.[24] The samples were refluxed in 2.0 ml of 0.5 N methanolic NaOH for 5 minutes. After cooling, 2.0 ml of 14% methanolic BF_3 was added and refluxing was repeated for 3 minutes. After cooling, 1.5 ml of water was added and methyl esters were extracted three times with 3.0 ml of hexane. The methyl esters, containing methyl heptadecanoate as an internal standard, were subjected to gas-liquid chromatography in a Varian Model 2860-10 instrument equipped with a polar column (600 by 0.4 cm, 10% of diethylene glycol succinate on Chromosorb W). The fatty acids were identified by their retention time relative to that of standard methyl ester mixtures. Fatty acid composition was quantitated by calculating peak area using the method of Carrol.[25] Erythrocyte lipid extractions were preceded by hemolysis and removal of essentially all residual hemoglobin. Cells were lysed in 30 volumes of 5 mM phosphate buffer, pH 8.0, and centrifuged at 22 x $10^3 g$ for 20 minutes. After centrifugation, the leucocyte button was aspirated away from the stroma and discarded. The ghosts were then repeatedly washed with 5 mM phosphate buffer (pH 8.0) and centrifuged until they appeared creamy white. Lipids were extracted as described by Folch,[26] followed by concentration *in vacuo*. The lipids were then re-extracted using the method of Bligh and Dyer,[27] the chlorofrom phase dried *in vacuo* and stored under N_2 at -5°C.

Pentose Phosphate Shunt Measurements

PPS measurements were performed using freshly drawn blood collected in heparinized Vacutainers (Becton-Dickinson Co., Rutherford, NJ). Washed cells diluted to 33% hematocrit using isotonic 5 mM glucose were exposed to ozone/oxygen or oxygen alone. 1-^{14}C-glucose or 2-^{14}C-glucose was added and the cell suspension was transferred to a 25-ml Erlenmeyer flask containing a center well into which a piece of filter paper saturated with Protosol (New England Nuclear, Boston, MA) was placed. After incubation, followed by acidification of the cell suspension with 1.0 ml 10% trichloracetic acid (TCA) to displace intracellular and extracellular CO_2 dissolved as bicarbonate, the filter paper strips were counted in Aquasol (New England Nuclear, Boston, MA).

Analytical Methods

Hemoglobin and reduced glutathione were quantitated employing techniques described in Beutler.[28] Hemoglobin standards used for spectrophotometer and reagent calibration were purchased from Sigma Chemical Co., St. Louis, MO. Protein was quantitated as described by Fairbanks *et al.*[29] Fluorescence measurements were made employing a Farrand MKI scanning spectrofluorometer. Osmotic fragility measurements were conducted using washed erythrocytes suspended in an isoosmotic pH 7.4 buffer containing 2.4 mM sodium barbital. A Fragiligraph Model D-3 (Kalmedic Instruments, New York) was used to generate osmotic fragility curves. Glucose was quantitated by the method of Dubois *et al.*[30] The method of Placer *et al.*[31] was used for TBA detection of malonaldehyde. Standardization was performed by generating malonaldehyde from its tetramethylacetal.

Enzyme Assays

Acetylcholinesterase (AChE) activity was measured using the method of Ellman *et al.*[32] where acetylthiocholine was the substrate. Glyceraldehyde-3-phosphate dehydrogenase (GPDH) was assayed according to Cori *et al.*[33] Glutathione peroxidase, glutathione reductase, 6-phosphogluconate dehydrogenase and glucose-6-phosphate dehydrogenase (G6PD) were all assayed using methods described by Beutler.[28] In the case of G6PD measurements, the enzyme activity was expressed in the absence of 6-phosphogluconate dehydrogenase activity.[34] In cases where intracellular enzymes of intact cells had to be measured, the cells were lysed with 35 μl of 0.1% sodium dodecyl sulfate per 3.0 ml cuvette volume, prior to substrate addition.

Vesicle Preparation

One-day-old to one-week-old human erythrocytes collected in citrate-phosphate-dextrose were used for vesicle preparation. Erythrocyte ghosts were prepared according to Fairbanks *et al.*,[29] which followed the principles of hypotonic lysis set forth by Dodge *et al.*[35] Right side out (RO) erythrocyte vesicles were prepared from ghosts allowed to reseal by warming in saline.[36,37] Inside out (IO) erythrocyte vesicles were prepared from ghosts employing a technique described by Steck and Kant,[38] which involves inducing endocytosis in fresh, well-washed ghosts.

RESULTS

The Effect of Ozone on Erythrocyte Membrane Integrity

Figures 1 and 2 demonstrate that ozone can strengthen the RBC membrane since there was protection from hypotonic hemolysis at a given salt concentration during the initial stages of ozone exposure. Higher concentrations of ozone then quickly reversed this effect and then disrupted membrane stability. Gassing RBC suspensions with oxygen for periods similar to ozone/oxygen exposure times resulted in no development of resistance to hypotonic hemolysis; rather, this control also had a disruptive effect on cellular integrity. This experiment was substantiated by the observation that similar ozone concentrations can cause RBC suspensions to become less susceptible to the lytic effects of 16 mM lyso-PC. Exposure to higher ozone concentrations then resulted in an enhanced rate of RBC hemolysis induced by lyso-PC (Data not shown).

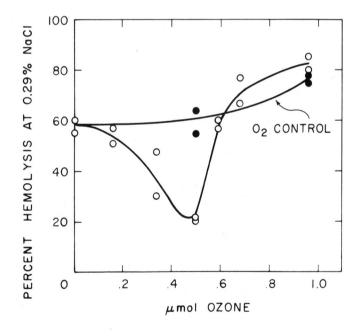

Figure 1. The effect of ozone on the hypotonic hemolysis properties of the human erythrocyte. Osmotic fragility curves were generated for erythrocytes and then a calculation was made for the percent hemolysis at 0.29% NaCl. Ozone delivery was 1.0 μmol/min. A 4.0-ml suspension of a 1:47 dilution of washed, packed RBC in NaCl-Na barbital buffer, pH 7.4, was exposed to ozone/oxygen at a flowrate of 20 ml/min. Erythrocytes were drawn immediately prior to start of experiment.

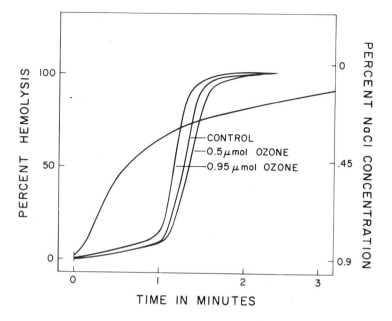

Figure 2. The osmotic fragility properties of ozonized human erythrocytes. Curves were generated as described in the section entitled Experimental Procedures. Data derived from osmotic fragility curves such as these were used in calculating Figure 1.

The Effects of Ozone on Erythrocyte Membrane Lipids

The lipids of the RBC membrane were examined to determine the extent of ozonolysis. Table I represents the phospholipid composition of controls and seven other RBC suspensions exposed to ozone for different lengths of time. In all these suspensions, gravimetric and chemical procedures showed that similar amounts of total lipid, cholesterol and phosphorus-containing lipids were extractable after lysis and washing of exposed intact cells. This eliminates the confusion that would result if ozonation of a membrane lipid changed its structure, thereby increasing its polarity and prohibiting extraction by organic solvents. There was no change in RBC phospholipid composition after exposure to massive amounts of ozone, and there was no increase in nonchromatographable lipid phosphorus at the origin of each TLC plate.

Table II shows that the fatty acid composition of the erythrocyte membrane was not demonstrably changed after ozone exposure. The values reported here for RBC phospholipid and fatty acid composition are similar to other published values.[39] In a second experiment using 0-11.25 μmol ozone, no changes in RBC phospholipid and fatty acid composition

Table I. Relative Phospholipid and Cholesterol Composition of Ozone-Exposed Human Erythrocytes[a]

Lipid[b]	0	μmol Ozone							O₂ Control
		9.5	19	28.5	38	47.5	57	66	
PE	29	31	30	29	32	28	29	32	30
PC	22	17	24	20	22	20	22	22	20
PS	17	18	17	17	18	21	18	18	19
SM	19	24	19	23	21	18	19	18	21
Origin	13	10	10	11	7	13	12	10	10
Chol/PL	0.55	0.55	0.57	0.46	0.54	0.56	0.55	0.51	0.53

[a]A 130-ml suspension of a 1:20 dilution of washed packed RBC in 0.154 M NaCl was exposed to ozone/oxygen at a flowrate of 20 ml/min. Ozone delivery was 9.5 μmol/min. Phosphate analyses were performed on individual lipids scraped from TLC plates. The TLC solvent system and lipid extraction techniques were as described in the Section entitled Experimental Procedures.
[b]Phospholipid composition expressed as mol % and the cholesterol to phospholipid ratio expressed was weight/weight.

Table II. Relative Fatty Acid Composition of Ozone-Exposed Human Erythrocytes[a]

Fatty Acid[b]	0	μmol Ozone							O₂ Control
		9.5	19	28.5	38	47.5	57	66	
16:0[c]	26	32	26	26	36	36	30	27	26
18:0	21	26	18	22	21	22	22	19	20
18:1	17	19	16	19	16	20	19	17	18
18:2	16	10	18	16	10	15	14	18	16
20:4	20	13	22	17	17	13	15	19	20

[a]Fatty acid methyl esters were prepared from the lipid extracts employed to obtain analyses in Table I.
[b]Fatty acid composition expressed as percentage of total peak area.
[c]The first number indicates chain length and the second indicates the number of double bonds.

were seen. Neither malonaldehyde nor Schiff's base fluorophore formed by the conjugation to two primary amine-containing phospholipids by malonaldehyde[40] could be observed in any of the lipid extracts. Some TBA positive material was observed on ozone-exposed RBC suspension supernatants or total hemolyzates, but the chromophore at 548 nm was felt to be contaminating heme released from hemoglobin into the TBA test reagent (*i.e.,* an indirect measure of cell lysis). Even though phospholipid and fatty acid composition were unchanged by ozone exposure, the RBC membranes were affected, evidenced by 12% hemolysis occurring in the 67-μmol ozone sample. AChE activity was 22% below control values and GPDH activity was 30% below control values. The inhibition of GPDH to an extent greater than the percentage of cell hemolysis suggests that ozone is affecting the cytoplasmic contents of intact erythrocytes.

Ozone was capable of oxidizing cholesterol at the Δ_5 double bond in a 1:1 stoichiometry, when cholesterol was dissolved in an organic medium (Figure 3). Ozonolysis was detected using a reagent specific for the sterol Δ_5[41] Although cholesterol is extremely sensitive to oxidation by ozone, Table I demonstrated that like fatty acids, cholesterol was not affected

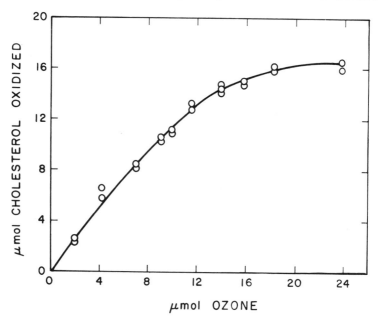

Figure 3. Reaction of ozone with the Δ_5 cholesterol double bond. Reaction mixtures consisted of 17 μmol cholesterol dissolved in 4.0 ml 95% ethanol. Ozone/oxygen was bubbled through each vessel at a flowrate of 20 ml/min with the ozone delivery being 10 μmol/min.

when positioned in the RBC membrane. The ratio of cholesterol to phospholipid remained constant in RBC suspensions exhaustively exposed to ozone.

The Permeability of Cell Membranes to Ozone

Several investigators have shown that erythrocyte AChE, located on the outside of the RBC membrane,[42] is inhibitable by ozone[43,44] and peroxides.[45] Figure 4a is consistent with these findings, showing an inhibition of RO RBC vesicle AChE by ozone. The residual AChE activity in IO vesicles was most likely due to contamination by RO vesicles or IO vesicles, which were permeable to both substrates and product. Figure 4b demonstrates that when ozone-exposed IO vesicles were lysed prior to AChE assays, restoration of enzymic activity occurred because of recovered enzyme accessibility to substrates. Even though most of the AChE was sequestered within the IO vesicle during gassing, following vesicle lysis there was an inhibition of AChE similar to the instance when AChE was present on the outside of the membrane during gassing. This suggests that ozone or some by-product of ozonolysis is capable of crossing a cell membrane and reacting inside a cell.

GPDH is located cytoplasmically in the human erythrocyte. Whether this enzyme is bound to the inner face of the cytoplasmic membrane or is a soluble cytoplasmic enzyme which binds to the membrane after lysis is still controversial.[46-48] GPDH, as noted in the previous section, was inhibited by ozone to an extent greater than AChE in intact erythrocytes. We have found that purified rabbit muscle GPDH is more susceptible to ozone than purified electric eel AChE. Figure 5a demonstrates that vesicle GPDH was inhibited to a greater extent than vesicle AChE. RO vesicles, which have GPDH internalized during gassing, showed no enzymic activity unless lysed prior to assay. There was then a measurable inhibition of GPDH by ozone which was greater than that which could be ascribed to the effect of the oxygen in the carrier gas. IO vesicles, which have GPDH exposed during gassing, were more susceptible to ozone than RO vesicle GPDH (lysed after gassing). Apparently, an enzyme surrounded by a membrane can be inhibited by ozone. Figure 5b illustrates that the lytic event had no effect on GPDH activity after ozone exposure.

Ozone Effects on RBC Glutathione Metabolism

When exposing cells from low to extremely high ozone concentrations, 6-phosphogluconate dehydrogenase was the only glutathione metabolism enzyme inhibited more than its oxygen control (Figure 6). No increases in glutathione metabolism enzyme-specific activities were observed as a response to oxidative stress.

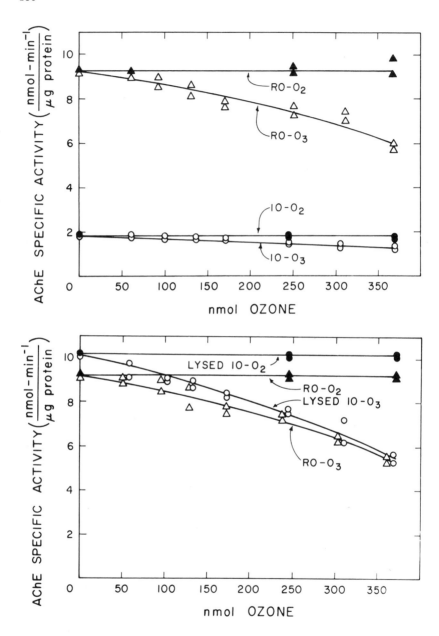

Figure 4.(a,b) Acetylcholinesterase activity of ozone-exposed IO and RO erythrocyte vesicles. 4.0 ml suspensions of 125 μg/ml vesicle protein were
exposed to 230 nmol/min ozone at a flowrate of 20 ml/min. IO
vesicles were lysed when noted by addition of 50 μl 2% sodium
dodecyl sulfate subsequent to exposure and prior to enzyme activity assays.

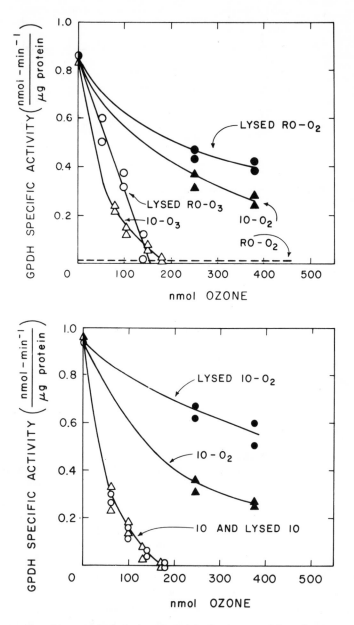

Figure 5.(a,b)   Glyceraldehyde-3-phosphate dehydrogenase activity of ozone-exposed IO and RO erythrocyte vesicles. The same vesicle suspensions described in Figure 4 were assayed for GPDH activity. IO and RO vesicles were lysed as described in Figure 4.

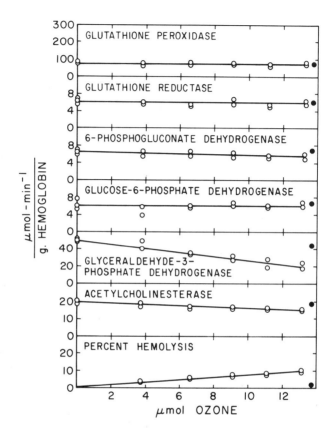

Figure 6. The effect of ozone on selected enzyme activities in intact human erythrocytes. Suspensions of 4.0 ml washed PRBC diluted 1/20 in 0.154 *M* NaCl were bubbled with 5 μmol/min ozone flowing in an oxygen carrier at a rate of 20 ml/min. Percent hemolysis was determined by quantitating the amount of extracellular Hb relative to the total RBC Hb present in the suspensions after gassing.

In cells that had been lysed prior to ozone exposure, all the glutathione metabolism enzymes except glutathione peroxidase were slightly inhibited by ozone (Figure 7). Lower amounts of marker enzyme inhibition occurred in this exposure, compared to Figure 6, perhaps a result of the release of oxidizable substances in the lysate, which were available to react to ozone. This observation also suggests that intracellular enzymes located away from the inner membrane face may be protected from ozone by reactive cytoplasmic materials.

Although there was no demonstrable effect of ozone on the specific activity of intact cell glutathione metabolism enzymes, exposure of intact

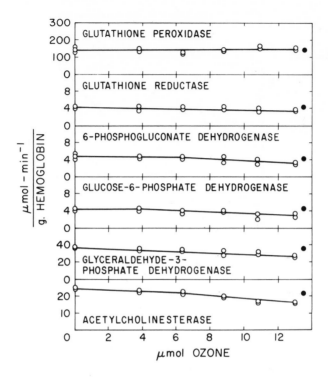

Figure 7. The effect of ozone on selected enzyme activities in lysed human erythrocytes. Packed RBC were lysed in 19 volumes distilled H_2O. Ozone/oxygen
was then bubbled through the lysate at 20 ml/min with an ozone delivery
rate of 5μmol/min.

cells to ozone enhanced the flow of glucose through the PPS (Figure 8).
$^{14}CO_2$ release from 1-^{14}C-glucose increased when cell suspensions were
treated with greater ozone concentrations. Oxygen controls showed no
effect. The slight increase in $^{14}CO_2$ release from 2-^{14}C-glucose suggests
that there is also a stimulation of pentose sugar recycling through the
shunt in ozone-exposed cells. Exposed cells gradually returned to the control rate of glucose metabolism (Figure 9). Higher concentrations of ozone
caused a greater stimulation of glucose metabolism which was followed
by a more rapid return to the control rate of metabolism.

Figure 10 presents evidence that there was oxidation of intracellular
GSH in ozone-exposed cells. Maximal GSH oxidation was measured immediately following exposure, implying that ozone directly oxidized GSH
to GSSG, as seen *in vitro*,[49] or to mixed disulfides between GSH and
cytoplasmic proteins. GSH levels then began to return to control values
via the reductive power provided by the stimulated PPS.

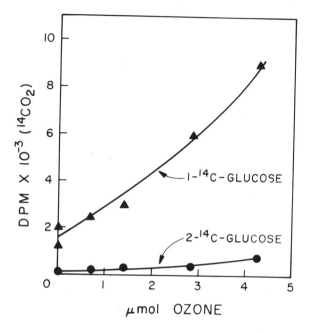

Figure 8. The influence of ozone on glucose routing through the pentose phosphate shunt. A suspension of washed PRBC diluted 1:3 in isotonic 5 mM glucose was preincubated at 25°C for three hours. Each 4.0 ml suspension contained 3 x 10^4 CPM/μmol glucose. 4.0 ml aliquots were exposed to ozone/oxygen flowing at 20 ml/min. Ozone delivery was 1.2 μmol/min. $^{14}CO_2$ evolution was measured in gassed suspensions, as described in the section entitled Experimental Procedures, following incubation for three hours.

DISCUSSION AND CONCLUSIONS

The decrease in the osmotic fragility of erythrocytes exposed to relatively low concentrations of ozone has not been reported before (Figures 1 and 2). Goldstein and Balchum[50] observed that 40 ppm of ozone ($\cong$ 0.8 μmol ozone) flowing through a 1:25 dilution of PRBC for two hours caused only an increase in RBC osmotic fragility, but they may have missed the transitory toughening effect. In our experiments, the initial ozone-induced stabilization observed could be due to cross-links formed between membrane components such as (1) ozone-oxidized protein sulfhydryls forming disulfide cross-links[49] between adjacent membrane proteins, or (2) ozone-produced oxidation of tryptophan to N-formyl kynurenine,[49,51] which then can form Schiff's base cross-links with vicinal primary amine-containing membrane components such as lysine,

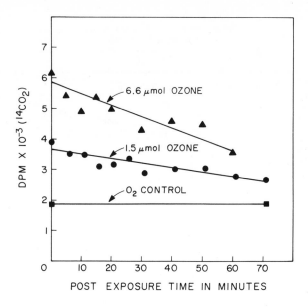

Figure 9. Routing of glucose through the pentose phosphate shunt subsequent to erythrocyte ozone exposure. Cells were exposed as described in Figure 8. $^{14}CO_2$ evolution rate was measured by initiating $^{14}CO_2$ trapping as much as 71 minutes after gassing and incubating suspensions for three hours.

glutamine, phosphatidyl ethanolamine and phosphatidyl serine. Preliminary experiments show that dithiothreitol-reducible cross-links in integral RBC membrane proteins can be caused by similar concentrations of ozone.

Tables I and II present evidence that substantial ozone damage (12% hemolysis and 20% to 30% enzyme inhibition) can occur in the absence of detectable fatty acid or cholesterol oxidation. From calculations using internal standards and published values,[35] it was estimated that in the 67-μmol ozone sample there was a 3.6-fold greater number of molecules of ozone reacting or degrading in the RBC suspension than fatty acid olefinic groups in the suspended cells. The consumption of ozone must be mostly by mechanisms other than fatty acid double bond oxidation. Another analysis[5] reported a 17% decrease in total fatty acids of ozonized red blood cells. The most marked decrease was in arachidonate. In these experiments, the RBC were gassed with enough ozone to correspond to about half the total number of membrane fatty acid olefinic groups. Our lipid analyses differ from the above data and demonstrate an absence of fatty acid oxidation, implying that the bulk of the ozone to which a cell is exposed reacts with protein and GSH.

Other investigators have reported no effect of ozone on membrane lipids.[12,52,53] Menzel et al.[54] reported decreases in lung oleic and

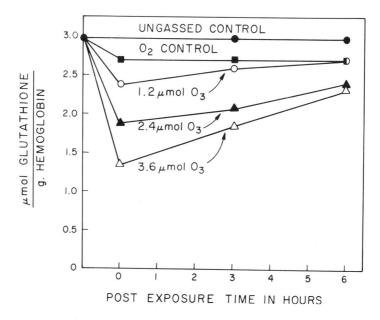

Figure 10. The effect of ozone on reduced glutathione content of ozone-exposed human erythrocytes. Cells were exposed as described in Figure 8. 0.4 ml aliquots of cells were assayed for GSH before, immediately after, three hours and six hours after exposure, as described in the section entitled Experimental Procedures.

linoleic acid and increases in arachidonic acid after exposure of rats to 1 ppm ozone for nine days. These results can be explained by alterations in lung lipid metabolism or an infiltration of arachidonate-rich serum as a result of the edemagenic nature of ozone.[55]

Teige *et al.*[44] observed little lysis of erythrocytes by ozone compared to the amount of lysis observed when ozonized lipid was added to cell suspensions. This suggests that ozone itself had little effect on RBC lipid. Goldstein *et al.*[51] observed ozone to oxidize tryptophan residues located on the outside of RBC; those buried within the lipid phase remained unaffected. The highly polar and charged ozone molecule may simply not react with olefins in the apolar region of the RBC membrane.

The resistance of erythrocyte fatty acids to ozonolysis does not rule out the possibility that in organelles such as mitochondria or microsomes ozone may be acting as a chaotrope. Ozone could destabilize the native membrane structure and allow lipid oxidation by molecular oxygen to occur such as seen when using SCN⁻, guanidine-HCl and other chaotropic agents.[56] Membranes containing closely associated cytochromes capable of catalyzing peroxidation reactions would then be especially susceptible to this phenomenon.

Ozone itself appears incapable of increasing the specific activities of RBC glutathione metabolism enzymes (Figure 6). Any effect of ozone on these enzymes would probably be inhibitory (Figure 7). Buckley et al.[3] observed human volunteers to experience decreased GSH levels and an increase in G6PDH-specific activity when breathing 0.5 ppm ozone for 2 3/4 hours. From their data it can be calculated that the volunteers would inhale 19.5 μmol ozone. Their measured loss of about 4 mg GSH/100 ml blood is equivalent to a total of 650 μmol of GSH oxidized—much more than can be accounted for by direct oxidation. The enucleated RBC is incapable of synthesizing G6PDH, so the increases in G6PDH activity observed would have to be explained in other ways than an adaptive response. It is possible that changes in this enzyme level could be due to the observed removal of cells from the circulation in persons breathing ozone.[57] Ozone-induced RBC sphering[58] would result in an increased rate of splenic sequestration,[59] thus selectively filtering out of the circulation older cells that are more susceptible to oxidant damage[60] and lowering the mean RBC age in the circulation. Younger cells are known to have an elevated G6PD activity.[61,62] Chow et al. have observed no changes in glutathione metabolism enzyme activities in the erythrocytes of rats and monkeys breathing 0.5 ppm ozone eight hours a day for seven days.[63]

The RBC are capable of altering their metabolism to recover from the oxidant stress presented by ozone (Figure 8). This metabolism proceeds at a rate dependent on the dosage of ozone received (Figure 9) and is capable of reversing large amounts of intracellular GSH oxidation (Figure 10). The stimulation of glucose flow through the PPS shown in Figures 8 and 9 could be explained by an elevated NADP/NADPH ratio activating G6PDH, the regulated first enzyme of this shunt. A 50% decrease in the NADP/NADPH ratio when glutathione reductase is actively using NADPH for GSSG reduction would double the G6PD activity;[62] the concentration of GSH alone seems to have no effect on the PPS.[60]

If substantial amounts of ozone can reach the circulation, these findings could have implications in the health of persons suffering from hematological disorders such as the common G6PD deficiency and rarer impairments of GSH metabolism. In addition, NADP/NADPH ratios are being studied in cells exposed to and recovering from ozone treatment to more fully understand ozone-induced changes in PPS regulation.

If ozone caused only the formation of GSSG, it can be calculated from Figure 10 that 6-10% of the ozone introduced to the cell suspension entered the cytoplasm and reacted with GSH.

From Figure 8, it can be calculated that a 4.0-ml suspension exposed to 4.0 μmol ozone metabolizes 0.22 μmol glucose through the PPS in three hours. This would provide the cells with 0.44 μmol NADPH and the capacity to produce 0.88 μmol of GSH from 0.44 μmol GSSG. From Figure 10 and knowing each cell suspension contains 0.36 g hemoglobin, it can be estimated that 4.0 μmol O_3 would oxidize 0.50 μmol GSH to 0.25 μmol GSSG or GSS-protein. Therefore, the erythrocytes have the metabolic capability to recover from the ozone-induced GSH oxidation even if mixed disulfides have to be reduced, in which case 1 NADPH would be required for every GSH produced. Knowing this, it appears that the recovery of GSH to control levels in Figure 10 is not proceeding at an optimum rate. This could be due to disulfide exchanges using endogenous GSH to reduce protein disulfides formed by ozone. GSH could also be required to metabolize normal or above-normal levels of H_2O_2. In any event, the GSH metabolism enzymes are operating far below their optimum specific activity measured under ideal conditions (480 μmol of GSH reduced/hr/4.0 ml cells) calculated from the rate-limiting G6PDH in Figure 6. This is most likely a result of the low substrate concentrations experienced within the cells. Intracellular GSH may never fully return to control levels unless *de novo* synthesis occurs, because erythrocytes contain an active transport mechanism for eliminating elevated and possibly cytotoxic levels of GSSG during periods of actue oxidative stress.[64]

It appears that ozone itself and possibly toxic by-products of ozonolysis are capable of affecting cell membranes (Figures 4a,b, 5a,b, 6 and 10). This knowledge will aid in determining mechanisms of ozone cytotoxicity. Experiments with alveolar macrophages and lung mitochondria isolated from ozone-exposed animals already implicate cytoplasmic enzyme inhibition as a factor leading to the inefficiency or demise of these cells and organelles.[65-67]

ACKNOWLEDGMENT

The authors thank Mr. Ted Kott of the San Bernardino Blood Bank for supplying blood. This research was supported by Research Grant ES00917 from the National Institute of Environmental Health Sciences.

REFERENCES

1. Niki, H., E. E. Daby and B. Weinstock. "Photochemical Smog and Ozone Reactions," *Adv. Chem.* 113:16-75 (1972).
2. Goldstein, B. P., B. Pearson, C. Lodi, R. D. Buckely and O. J. Balchum. *Arch. Environ. Health* 16:648-650 (1968).
3. Buckley, R. D., J. D. Hackney, K. Clark and C. Posin. *Arch. Environ. Health* 30:40-43 (1975).

4. Gardner, D. E., J. W. Illing, F. J. Miller and D. L. Coffin. *Res. Commun. Chem. Path. Pharm.* 9:689-696 (1974).
5. Balchum, O. J., J. S. O'Brien and B. D. Goldstein. *Arch. Environ. Health* 22:32-34 (1971).
6. Roehm. J. N., J. G. Hadley and D. B. Menzel. *Arch. Environ. Health* 23:142-148 (1971).
7. Menzel, D. B., R. J. Slaughter, A. M. Bryant and H. O. Juaregi. *Arch. Environ. Health* 30:234-236 (1975).
8. Menzel, D. B., R. J. Slaughter, A. M. Bryant and H. O. Juaregi. *Arch. Environ. Health* 30:296-301 (1975).
9. Goldstein, D. B., C. Lodi, C. Collinson and O. J. Balchum. *Arch. Environ. Health* 18:631-635 (1969).
10. Teige, B., T. T. McManus and J. B. Mudd. *Chem. Phys. Lipids* 12: 153-171 (1974).
11. Mudd, J. B., T. T. McManus, A. Ongun and T. E. McCullough. *Plant Physiol.* 48:335-339 (1971).
12. Swanson, E. S., W. W. Thomson and J. B. Mudd. *Can. J. Bot.* 51: 1213-1219 (1973).
13. Mudd, J. B., T. T. McManus, A. Ongun and T. E. McCullough. in *Proc. 2nd Intl. Clean Air Cong.* (New York: Academic Press, 1971), pp. 256-260.
14. Chow, C. K., and A. L. Tappel. *Lipids* 7:518-524 (1972).
15. Chow, C. K., C. J. Dillard and A. L. Tappel. *Environ. Res.* 7:311-319 (1974).
16. Chow, C. K., M. G. Mustafa, C. E. Cross and B. K. Tarkington. *Environ. Physiol. Biochem.* 5:142-146 (1975).
17. Pierce, T. H., G. K. York, C. E. Franti and C. E. Cross. *Arch. Environ. Health* 31:290-293 (1976).
18. Freeman, G., R. J. Stephens, D. L. Coffin and J. F. Stara. *Arch. Environ. Health* 26:209-216 (1973).
19. Salzmann, B., and N. Gilbert. *Anal. Chem.* 31:1914-1920 (1959).
20. Hendricks, R. H., and L. B. Larsen. *Am. Ind. Hyg. Assoc. J.* 27: 80-84 (1966).
21. Nichols, B. W., and A. T. James. *Fette Seifen Anstrich.* 66:1003-1007 (1964).
22. Ames, B. N. in *Methods in Enzymology,* Vol. 8, B. O'Malley, Ed. (New York: Academic Press, 1966), pp. 115-118.
23. Rudel, L., and M. Morris. *J. Lipid Res.* 14:364-366 (1973).
24. Beare-Rogers, J. L. *Can. J. Biochem.* 47:257 (1969).
25. Carroll, K. K. *Nature* (London) 191:377-378 (1961).
26. Folch, J., M. Lees and G. H. Sloane-Stanley. *J. Biol. Chem.* 226: 497-501 (1957).
27. Bligh, E. G., and W. J. Dyer. *Can. J. Biochem. Physiol.* 37:911-917 (1959).
28. Beutler, E. *Red Cell Metabolism,* A Manual of Biochemical Methods. *(New York: Grune and Stratton, 1975), pp. 11,12.*
29. Fairbanks, G., T. L. Steck and D. F. H. Wallach. *Biochemistry* 10: 2606-2616 (1971).
30. Dubois, M., K. A. Gilles, J. K. Hamilton, P. A. Rebers and F. Smith. *Anal. Chem.* 28:350-356 (1956).

31. Placer, Z. A., L. L. Cushman and B. C. Johnson. *Anal. Biochem.* 16:359-366 (1966).
32. Ellman, G. L., K. D. Courtney, V. Andres, Jr. and R. M. Featherstone. *Biochem. Pharmacol.* 7:88-95 (1961).
33. Cori, G. T., M. T. Slei and C. G. Cori. *J. Biol. Chem.* 173:605-612 (1948).
34. Glock, G. E., and P. McLean. *Biochem. J.* 55:400-408 (1953).
35. Dodge, J. T., and G. B. Phillips. *J. Lipid Res.* 8:667-676 (1967).
36. Bodemann, H., and H. Passow. *J. Membr. Biol.* 8:1-26 (1972).
37. Johnson, R. M. *J. Membrane Biol.* 22:231-253 (1975).
38. Steck, T., and J. A. Kant. in *Methods in Enzymology,* Vol. 31, Par Part A, S. Fleischer and L. Packer, Eds. (New York: Academic Press, 1974), pp. 173-175.
39. Dodge, J. T., and G. B. Phillips. *J. Lipid Res.* 8:667-675 (1967).
40. Trombly, R., A. L. Tappel, J. G. Coniglio, W. M. Grogan, Jr., and R. K. Rhamy. *Lipids 10*:591-596 (1975).
41. Zlatkis, A., and B. Zak. *Anal. Biochem.* 29:143-148 (1969).
42. Bellhorn, M., O. O. Blumenfeld and P. M. Gallop. *Biochem. Biophys. Res. Commun.* 39:267-273 (1970).
43. Goldstein, B. D., B. Pearson, C. Lodi, R. D. Buckely and O. J. Balchum. *Arch. Environ. Health* 16:648-650 (1968).
44. P'an, A. Y. S., and Z. Jegier. *Arch. Environ. Health* 21:498-501 (1970).
45. O'Malley, B., C. Mengel, W. Merriwether and L. Zirkle. *Biochemistry* 5:40-47 (1966).
46. Shin, B. C., and K. L. Carraway. *J. Biol. Chem.* 248:1436-1444 (1973).
47. Tillman, W., A. Cordua and W. Schroter. *Biochim. Biophys. Acta* 382:157-171 (1975).
48. Maretzki, D., J. Groth, A. G. Tsamaloukas, M. Grundel, S. Kruger and S. Rappaport. *FEBS Lett.* 39:83-87 (1974).
49. Mudd, J. B., R. Leavitt, A. Ongun and T. T. McManus. *Atmos. Environ.* 3:669-682 (1969).
50. Goldstein, B. D., and O. J. Balchum. *Proc. Soc. Exp. Biol. Med.* 126:356-358 (1967).
51. Goldstein, B. D., and E. M. McDonagh. *Environ. Res.* 9:179-186 (1975).
52. Rich, S., and H. Tomlinson. in *Air Pollution Effects on Plant Growth,* W. M. Dugger, Ed., A.C.S. Symposium Series 3 (1974), pp. 76-82.
53. Dowell, A. R., L. A. Lohrbauer, D. Hurst and S. D. Lee. *Arch. Environ. Health* 21:121-127 (1970).
54. Menzel, D. B., J. B. Roehm and S. D. Lee. *J. Agr. Food Chem.* 20:481-487 (1972).
55. Kyei-Aboagye, K., M. Hazucha, I. Wyszogrodski, D. Rubenstein and M. E. Avery. *Biochem. Biophys. Res. Commun.* 54:907-913 (1973).
56. Hatefi, Y., and W. G. Hanstein. *Arch. Biochem. Biophys.* 138:73-86 (1970).
57. Buckley, R., J. Hackney, K. Clark and C. Posin. *Fed. Proc.* 36:630 (1977).
58. Lamberts, H., R. Brinkman and T. Veninga. *Lancet* (January 18, 1964).

59. Walls, R., S. K. Kumar and P. Hochstein. *Arch. Biochem. Biophys.* 174:463-468 (1976).
60. Eaton, J. W., and G. J. Brewer. in *The Red Blood Cell,* Vol. 1, D. M. Surgenor, Ed. (New York: Academic Press, 1974), pp. 435-471.
61. Cohen, N. S., J. E. Ekholm, M. G. Luthra and D. J. Hanahan. *Biochim. Biophys. Acta.* 419:229-237 (1976).
62. Yoshida, A. *Science* 179:532-537 (1973).
63. Chow, C. K., M. G. Mustafa, C. E. Cross and B. K. Tarkington. *Environ. Physiol. Biochem.* 5:142-146 (1975).
64. Prchal, J., S. K. Srivastava and E. Beutler. *Blood* 46:111-117 (1975).
65. Goldstein, E., W. Lippert and D. Warshauer. *J. Clin. Invest.* 54:519-526 (1974).
66. Warshauer, D., E. Goldstein, P. D. Hoeprich and W. Lippert. *J. Lab. Clin. Med.* 83:228-235 (1974).
67. Mustafa, M. G. *Arch. Biochem. Biophys.* 162:585-592 (1974).

CHAPTER 10

PULMONARY RESPONSES TO
SULFURIC ACID AEROSOLS

L. W. Schwartz, Y. C. Zee, B. K. Tarkington,
P. F. Moore and J. W. Osebold

Departments of Pathology and Microbiology
School of Veterinary Medicine and
  California Primate Research Center
University of California
Davis, California 95616

INTRODUCTION

Altshuller has recently summarized air surveillance measurements of sulfur dioxide and suspended water-soluble sulfates.[1] His summary indicates that sulfate-containing aerosols are broadly distributed throughout large regions of the eastern and midwestern United States. Health effects of sulfur-containing compounds are of increasing concern because the incidence of acute respiratory disease has been associated with environments heavily polluted with sulfur dioxide and suspended sulfates,[2,3] and because the present energy situation will require increased use of fossil fuels containing high concentrations of sulfur.

Morphological effects of sulfuric acid aerosols on the respiratory tract of guinea pigs, mice, rats and monkeys have been reported by our laboratory.[4] Particular emphasis was placed on the use of acid droplets in the submicron range and the use of precise morphological techniques to provide a critical evaluation of multiple sites from the respiratory tract. Exposure periods ranged from 4-14 days to aerosols of H_2SO_4 having a mass median aerodynamic diameter (MMAD) of 0.3-0.6 μm. Levels of sulfuric acid required to produce morphological effects in guinea pigs during these short-term exposures were 60-70 mg/m^3.

Lesions were observed in mice following exposure to 125 mg/m^3. Damage to the respiratory system of guinea pigs consisted of microscopic areas of erosion and ulceration in bronchi and bronchioles. Coagulation necrosis of bronchial epithelium, smooth muscle and cartilage, plus edema and hemorrhage within alveoli were observed. Airway changes in mice were similar in nature to those observed in guinea pigs, but the distribution differed in that lesions were limited to the larynx and adjacent trachea. Morphological damage to intrapulmonary airways or pulmonary parenchyma was not observed in mice.

These concentrations of sulfuric acid necessary to produce pulmonary damage following short-term exposure are certainly excessive when compared to high ambient levels of 20-25 μg/m^3, which occur during severe episodes of industrial smog; but perhaps of most interest was the observation that structural change did not occur in the nonhuman primate lung following short-term exposure to mass concentrations as high as 500 mg/m^3. Rats also appeared to be a resistant species as structural changes were not observed following continuous exposure to 172 mg/m^3. for seven days. This variability of response between four laboratory animal species reemphasizes the necessity to evaluate toxicological effects in multiple-animal species to be able to make a meaningful extrapolation of the data to human exposures.

The major purpose of this chapter is to expand our observations on mice. The systemic effects of sulfuric acid exposure on mice were monitored using several hematological parameters and urinalysis since sulfates may exert a systemic influence following absorption from the aqueous surface of the respiratory tract. Selected changes in defense systems of the mouse respiratory tract were evaluated by determining the ability of both tracheal organ cultures and alveolar macrophages to produce interferon, since production of this cell product serves as a sensitive indicator of altered cell function and its release has implications relative to the resistance of the host to viral infections.

An additional objective is to present the results of an LD$_{50}$ study in specific pathogen-free (SPF) guinea pigs using our system of generating submicron droplets of sulfuric acid, since previous studies have demonstrated altered responses by changing the acid droplet size.

METHODS AND EXPERIMENTAL PROCEDURES

Sulfuric Acid Generation and Monitoring

The sulfuric acid generation system and the procedures for determining the MMAD of the acid droplets and sulfate concentration have been described in detail elsewhere.[4,5] Exposure chamber conditions were essentially the same as those described previously.[4] The details of each exposure are described in Table I, II and III.

Animals

Male Swiss-Webster mice 60±3.5 days of age, free of chronic respiratory diseases (Hilltop Laboratory Animals, Inc., Chatsworth, PA) were housed 6-8/cage under microbiologic filters prior to exposure but during exposure were housed within stainless steel wire mesh cages.

Table I. Mouse Exposure Data Summary

	Number of Mice	MMAD[a] (μm)	σ	Mass Concentration (mg/m^3)	Exposure Length (days)
Exposure I	80	0.32	1.4	125	14
Exposure II	114	0.45	1.6	141	14
Exposure III	175	0.62	1.7	154	10

[a]Mass median aerodynamic diameter.

Table II. Spontaneous Death Losses During Exposure[a]

	Exposure I		Exposure II		Exposure III	
Day of Exposure	No. Dead	Percent Dead[b]	No. Dead	Percent Dead	No. Dead	Percent Dead
1	0	0	0	0	13	7.43
2	8	11.94	0	0	23	16.55
3	2	3.33	1	0.88	9	7.14
4	3	5.77	0	0	9	10.59
5	4	9.30	2	2.22	11	14.86
6	8	26.67	0	0	4	6.25
7	0	0	3	3.45	4	6.78
8	0	0	0	0	2	6.90
9	1	4.17	0	0	4	16.00
10	2	10.53	1	1.69	1	4.00
11	1	5.56	0	0	–	–
12	1	7.69	2	3.51	–	–
13	0	0	0	0	–	–
14	0	0	5	9.61	–	–

[a]See Table I for exposure details.
[b]Calculated as percent dead of remaining population on each sampling day.

Table III. Sulfuric Acid Aerosol LD_{50} Determination of Guinea Pigs Using the Reed-Muench Method

Mass Concentration (mg/m³)	MMAD (μm)	σ	No. of Guinea Pigs	No. Dead	Accumulated No. Dead	Accumulated No. Alive	Accumulated Total	Cumulated Percent Mortality
38	0.40	2.1	10	0	0	34	34	0
47	0.40	2.2	10	0	0	24	24	0
62	0.32	1.6	10	1	1	14	15	6.67
105	0.33	1.5	10	5	6	5	11	54.55
220	a	a	10	10[b]	16	0	16	100.00

[a] Length of exposure was not of adequate length to make these determinations.
[b] Six guinea pigs died within three hours of start of exposure and the remaining four were killed in extreme respiratory distress to collect fresh tissues for morphological evaluation. Calculations made with the assumption that all animals would have died.

$LD_{50} = 100$ mg/m³, 95% confidence limits are 76 - 132 mg/m³.

Hartley albino female guinea pigs purchased as SPF animals (Charles Rivers, Wilmington, MA) were used at an age of 51-65 days and mean body weight of 410 g. The exposure proceeded continuously for 7 days. The LD_{50} determination was calculated using the Reed-Muench method.[6]

Tissue Collection for Morphology

Tissues were collected from five mice on each of days, 1, 2, 3, 4, 5, 7, 9, 11 and 14 of exposure I only. Methods of lung fixation and evaluation have been described previously in detail.[7] Brain, liver, kidneys, spleen, pancreas, stomach and skin were collected and fixed by submerging in Karnovsky's fixative and examined by light microscopy following routine processing.

Hematology and Clinical Chemistry

Blood and urine samples were obtained at autopsy. Values were obtained using routine hematological and clinical chemistry procedures.

Interferon Assay

Viruses. The California 11914 strain of Newcastle disease virus (NDV) was used as an interferon inducer. The Indiana strain of vesicular stomatitis virus (VSV) was employed in the assay.

Polynucleotide. Poly I:C (inosine:cytidine) was purchased from Biopolymers, Chagrin Falls, OH.

Tracheal Organ Cultures. Organ cultures were prepared according to the technique described by Hoorn and Tyrrell[8] and previously used by our laboratory.[9] Tracheal explants from two mice were placed in a sterile 35 mm x 10 mm plastic culture dish to which 5 ml of maintenance medium (medium 199) containing 50 µg/ml gentamicin and 0.3% bovine serum albumin were added. The cultures were incubated at 35°C in a humidified atmosphere of 5% CO_2 in air.

Alveolar Macrophage Cultures. Alveolar macrophages were obtained from mice using a slight modification of the technique described by Medin *et al.*[10] Lung lavages from 18 mice were pooled, and the cells sedimented by centrifugation at 121 g for 30 min at 5°C. The supernatant was decanted and the cells resuspended in 4 ml of MEM containing 10% fetal calf serum and 50 µg/ml gentamicin. The cell suspension was then dispensed equally into 10 x 35-mm plastic culture dishes in 1-ml amounts. The macrophages were allowed to adhere for a 2 hr period at 37°C in a 5% CO_2 atmosphere. Unattached cells were removed by aspirating, and adherent cells were washed three times with 2 ml of Hank's solution. Cell counts from 20 petri dishes indicated that each culture dish contained an average of 4.0 x 10^5 (2.5-5.0 x 10^5) adherent cells.

Interferon Production in Organ Culture. Organ cultures were inoculated with interferon inducers on the fourth hour of incubation as follows: the medium was removed and 0.2 ml of NDV suspension containing 5 x 10^5 plague-forming units (pfu) or 0.2 ml of MEM containing 10 μg of poly I:C was dripped onto the tissue fragments. Control organ cultures received 0.2 ml of MEM. Adsorption of the inducers proceeded in the cultures for 1 hr while incubating at 35°C in 5% CO_2 environment. The cultures were then washed three times with 2 ml of medium 199 and replenished with 5 ml of medium 199 containing 0.3% bovine serum albumin and 50 μg/ml of gentamicin. Tracheal cultures were incubated for 16 hr at 35°C in a humidified 5% CO_2 atmosphere. At the end of this incubation period, the culture fluid was harvested by aspiration and stored at -20°C.

Interferon Production in Macrophage Cultures. The monolayers of alveolar macrophages were induced to produce interferon using an inoculum of NDV at a multiplicity of infection of 10 or 1 ml of MEM containing 50 μg of poly I:C. Control cultures received 1 ml of MEM. After absorption for 1 hr, the cell cultures were washed three times with 2 ml of MEM after which 5 ml of MEM containing 50 μg/ml of gentamicin was added. After incubation for 16 hr the culture fluid was collected and stored at -20°C.

Interferon Assay. The culture fluid from the organ cultures and macrophage cultures was assayed for interferon using the plaque reduction technique on L-cells with VSV as the challenge virus. The culture fluids were adjusted to pH 2.0 with 2 *M* perchloric acid and kept at 4°C for 4 days. The resulting precipitate was removed by centrifugation at 80,000 g for 1 hr, and the pH of this supernatant readjusted to pH 7.0 with 2 *M* NaOH. The interferon assay was then carried out on this supernatant using L-cells and VSV as the challenge virus.

The interferon titer was expressed as the reciprocal of the highest dilution of the sample showing 50% plaque reduction. A standard reference mouse interferon (catalog number G002-90A-511) provided by the National Institutes of Health was included in each assay. One unit of interferon in our assay was equivalent to two units of the reference interferon.

The antiviral substance demonstrated in this study was considered to be mouse interferon for the following reasons. It was shown to be (1) stable at pH 2.0, (2) nonsedimentable at 100,000 g for 1 hr, (3) inactivated by trypsin treatment, (4) specific for host species, (5) inhibited by Actinomycin D, and (6) inactivated at 56°C for 1 hr.[11]

RESULTS

The pattern of spontaneous deaths during the three exposures of mice has been summarized in Table II. Morphological changes were as described previously[4] and limited to the larynx and upper trachea. The larynx was examined from 41 of the 45 exposed mice, and lesions of similar nature were observed in 27 mice, whereas none of the 15 control mice demonstrated evidence of laryngeal damage. Microscopic multifocal areas of acute coagulation necrosis were observed in the liver of two of five mice sampled on the second day of exposure. Noninflammatory response was associated with this necrosis. This hepatic change was not observed in control mice or in other exposed groups except for one of the mice dying spontaneously on the third day of exposure. The pathogenesis or significance of these focal areas of hepatic necrosis was unknown.

Additional changes, which did not appear related to exposure nor significant to the health of the mouse, included focal peribronchiolar and perivascular lymphoid clusters in the lungs of eleven mice (ten exposed, one control), focal interstitial lymphoid clusters in the kidney of nine mice (exposed), and six mice (exposed) had cutaneous acariasis. Lesions were not recognized in sections of stomach, spleen, pancreas, or brain.

Hematological results are reported in Table IV. The elevated white cell count, packed cell volume and plasma protein on days two and three most likely reflect a degree of dehydration and hemoconcentration. The total while cell count on blood indicates a trend of decreasing numbers of circulating white cells as exposure length increased and the differential cell counts indicated a neutropenia and correlated lymphophilia as exposure progressed.

The urinalysis results are reported in Table V. The elevated specific gravity early in the exposure may again reflect dehydration. Significant changes in the parameters examined were not observed. In addition, acetone, sugar or crystals within the urine were never observed.

Results of the interferon production assay on alveolar macrophage cultures and tracheal organ cultures are reported in Tables VI and VII, respectively. After induction by either poly I:C or NDV, the interferon titers of macrophage cultures from mice exposed to sulfuric acid are, as a whole, lower than those of control cultures. Results of one experiment on tracheal organ cultures did not demonstrate a difference between controls and exposed mice.

Exposure details and results of the LD_{50} study on SPF guinea pigs are reported in Table III. The LD_{50} was calculated to be 100 mg/m^3 using sulfuric acid aerosol droplets of 0.3-0.4 μm. The 95% confidence limits were determined to be 76-132 mg/m^3.

Table IV. Hematology Results

	Exposed Mice[a]/Exposure Day							Control Mice[b]
	1	2	3	4	9	11	14	
WBC/μl mean	3,700	4,240	5,580	3,100	4,560	2,840	3,125	4,580
range	3,000-4,500	2,300-6,300	4,200-7,800	2,400-4,500	2,000-6,700	2,200-3,800	1,400-5,200	2,600-7,900
PCV % mean	37.8	32.6	44.6	42.0	35.4	34.4	33.2	33.6
range	33-41	37-41	41-51	34-48	31-38	28-40	29-41	22-41
Plasma Proteins, gm/dl mean	4.8	5.0	5.0	4.6	3.5	3.2	3.4	4.0
range	4.5-5.0	4.7-5.3	4.2-5.4	3.8-5.7	3.1-4.1	2.6-3.9	2.5-4.7	3.0-5.1
Plasma Fibrinogen, mg/dl mean	200	200	187	133	160	220	133.3	270
range	100-300	100-400	50-400	100-200	100-200	200-300	100-200	100-500
Differential[c]								
Bands/mean	0	59	0	0	0	210	150	0
Neutrophils	7,428	2,459	1,362	1,147	1,414	721	487	1,791
Lymphocytes	2,087	1,518	3,895	1,931	2,982	1,505	2,275	2,606
Eosinophils	74	102	100	20	36	34	81	0
Monocytes	111	102	223	0	128	62	69	0

[a] Each analysis determined on an average of values from five mice per day.
[b] An average of values from 15 mice tested on three separate days.
[c] Absolute value.

Table V. Urinalysis Results

	Exposed Mice[a]/Exposure Day								Control Mice[b]
	1	2	3	4	7	9	11	14	
Specific Gravity	1.093	>1.070	>1.035	1.039	ND[c]	>1.070	>1.035	>1.070	>1.035
pH	6.5	6.0	7.0	6.0	6.0	6.5	6.0	6.5	6.8
Protein	1+	2+	ND	1+	Neg	1+	±	ND	Trace to 1+
Casts[d]	NS[e]	NS	0.3	NS	NS	NS	NS	NS	NS
WBC[d]	NS	Rare	NS	1-2	NS	Rare	NS	0-1	0-5
RBC[d]	4-7	1-3	4-10	TNTC[f]	Rare	Rare	40-50	16-18	ND
Occult Blood in Supernatant	ND	Neg	Neg	2+	Neg	Neg	3+	Neg	Neg

[a] Determined on a pooled sample of 5 mice/day.
[b] Average values from 15 mice tested on 3 separate days.
[c] Not Done.
[d] Number/high power field.
[e] None seen.
[f] Too numerous to count.

Table VI. Interferon Production by Alveolar Macrophage Cultures from Mice[a]

		Interferon Titer (units/4 ml)			
		Sulfuric Acid-Exposed Mice		Control Mice	
Exposure Day	Interferon Inducer	Exposure II	Exposure III	Exposure II	Exposure III
0	None			0	0
	NDV			512	256
	Poly I:C			512	256
3	None	0	0		0
	NDV	64	64		128
	Poly I:C	8	128		256
7	None	0	0		
	NDV	16	32		
	Poly I:C	8	64		
10	None		0		0
	NDV		32		128
	Poly I:C		128		312
14	None	0		0	
	NDV	32		512	
	Poly I:C	8		256	

[a] Macrophage cultures were prepared from lung lavages of 18 mice. Macrophage cultures were inoculated either with 50 μg of poly I:C (inosine:cytidine) or Newcastle disease virus (NDV) at a multiplicity of 10. Culture fluids were collected and processed for interferon titration as described under Methods and Experimental Procedures.

DISCUSSION AND CONCLUSIONS

We have previously provided initial information relative to pulmonary morphological changes in the guinea pig, mouse, rat and rhesus monkey following short-term, high-level exposures to submicron droplets of sulfuric acid mist.[4] Our experience indicated that the guinea pig and mouse were the more sensitive species. This study expands the information relative to these two species.

Our interest in interferon production was threefold. First, as previously discussed, an increased incidence of acute respiratory disease has been associated with environments heavily polluted with suspended sulfates[2,3] and production of interferon relates to resistance against viral infections. Second, interferon is a protein synthesized and released by cells when induced by viruses or polynucleotides, and its production and release serves as a sensitive indicator of cell function. Third, in view of our recent findings that tracheal organ cultures from mice exposed to 0.8 ppm ozone

Table VII. Interferon Production by Tracheal Organ Cultures from Mice[a]

Exposure Day	Interferon Inducer	Interferon Titer (units/4 ml)	
		Sulfuric Acid-Exposed Mice	Control Mice
		Exposure II	Exposure II
0	None		0
	NDV		128
	Poly I:C		64
3	None	0	
	NDV	32	
	Poly I:C	64	
7	None	0	
	NDV	32	
	Poly I:C	64	
14	None	0	0
	NDV	8	64
	Poly I:C	64	128

[a]Tracheal organ cultures from two mice were inoculated with either 10 μg of poly I:C (inosine:cytidine) or 5×10^5 pfu of Newcastle disease virus (NDV). After incubation at 35°C for 16 hours, culture fluids were collected and processed for interferon titration as described under Methods and Experimental Procedures.

for 11 days lost the capacity to produce interferon upon induction *in vitro*,[9] and the observation by Valand *et al.*[12] that the alveolar macrophage from rabbits exposed to nitrogen dioxide were not able to produce interferon when challenged with parainfluenza virus *in vitro*, the effect of sulfuric acid aerosols on interferon production was of interest. Interferon within culture fluid of exposed alveolar macrophages was reduced as compared to unexposed controls but this reduction may reflect either a diminished ability of these cells to produce this protein product or an altered capacity to release it. Certainly the mechanism(s) involved in reducing the amount in culture fluids from sulfuric acid-exposed alveolar macrophages remains unknown. Possible mechanisms involved related to the effects of other inhaled pollutants have been speculated on previously.[9,12] Interferon plays an important role in viral infections and often is present prior to antibody formation, as one of the initial host defense processes. The above evidence indicates that inhaled pollutants can alter this defense effort, and additional investigation to provide more precise information is required to relate these changes to increased susceptibility to respiratory tract infections during episodes of air pollution.

Analysis of blood and urine failed to demonstrate definite direct systemic effects of inhaled sulfates. Initial evidence of dehydration was present which most likely reflects the failure of mice to consume adequate amounts of water early in exposure. There was a definite trend of a neutropenia and lymphophilia; however, an explanation for this was not evident. It seems unlikely that the laryngeal lesion could in itself produce the neutropenia. Light microscopic examination of major organ systems failed to demonstrate exposure-related lesions in tissues other than the larynx and trachea.

Pattle et al.[13] observed that as the particle size decreased, the mass concentration necessary to produce an LD_{50} increased. Their early studies using acid droplets of 2.7 μm produced an LD_{50} in 200-250-g guinea pigs at a mass concentration of 27 mg/m^3, but acid droplets of 0.8 μm required a mass concentration of 60 mg/m^3 to produce an LD_{50}. Amdur, in a series of studies including both animals and man, has examined effects of sulfuric acid mist on structural and functional characteristics of the respiratory system.[14-19] In a review of the toxic effects of sulfate aerosols, she specifically emphasized that particle size was extremely important in producing pulmonary effects. Particles in the submicron range produced the greatest alteration in airflow resistance. Amdur indicated that a particle size of 0.8 μm was the most effective in increasing airflow resistance when 7 μm, 2.5 μm and 0.8 μm droplets were compared.[14] We are unaware of studies, other than ours, that have concentrated on species sensitivity and pulmonary effects after exposure to acid droplets in the 0.3-0.4 μm range. Generation of an aerosol of this size range was made possible by the development of the previously described acid generation system.[5] The LD_{50} for SPF guinea pigs 51-65 days old and having an average body weight of 410 g exposed to sulfuric acid droplets in the submicron particle range of 0.3-0.4 μm was 100 mg/m^3. This observation is consistent with the concept proposed by Pattle et al.[13]

Our results indicate that the respiratory tract is well shielded against structural damage induced by short-term exposure to sulfuric acid aerosols. Functional changes will occur at levels below those used in this study and before irreversible structural changes occur. Reduced amounts of interferon in alveolar macrophage cultures reflect functional alterations in the deep lung with an absence of structural damage. It is important to emphasize that these findings following short-term, high-level exposures should not, and cannot, be used to predict the effects of the more realistic long-term, low-level types of exposure. More importantly, it is necessary to consider the interactions of various gaseous and particulate environmental pollutants, because synergy may amplify the pulmonary effects.

ACKNOWLEDGMENTS

This study was supported by Contract No. 68-02-1732 from the Environmental Protection Agency and Grant No. RR00169 from the National Institutes of Health.

REFERENCES

1. Altshuller, A. P. "Atmospheric Sulfur Dioxide and Sulfate Distribution of Concentration at Urban and Nonurban Sites in United States," *Environ. Sci. Technol.* 7(8):709 (1973).
2. French, J. G., G. Lowrimore, W. C. Nelson, J. F. Finklea, T. English and M. Hertz. "The Effect of Sulfur Dioxide and Suspended Sulfates on Acute Respiratory Disease," *Arch. Environ. Health* 27:129 (1973).
3. Shy, C. M., and J. F. Finklea. "Air Pollution Affects Community Health," *Environ. Sci. Technol.* 7(3):204 (1973).
4. Schwartz, L. W., P. F. Moore, D. P. Chang, B. K. Tarkington, D. L. Dungworth and W. S. Tyler. "Short-term Effects of Sulfuric Acid Aerosols on the Respiratory Tract. A Morphological Study in Guinea Pigs, Mice, Rats, and Monkeys," In *Biochemical Effects of Environmental Pollutants*, S. D. Lee, Ed. (Ann Arbor, MI: Ann Arbor Science Publishers, Inc., 1977), pp. 257-271.
5. Chang, D. P. Y., and B. K. Tarkington. "Experience with a High Output Sulfuric Acid Aerosol Generator," *J. Am. Ind. Hyg. Assoc.* 38:493-497 (1977).
6. Woolf, C. M. *Principles of Biometry* (Princeton, NJ: D. Van Nostrand, Co., 1968), pp. 293-296.
7. Schwartz, L. W., D. L. Dungworth, M. G. Mustafa, B. K. Tarkington and W. S. Tyler. "Pulmonary Responses of Rats to Ambient Levels of Ozone," *Lab. Invest.* 34(6):565 (1976).
8. Hoorn, B., and H. J. Tyrrell. "On the Growth of Certain "Newer" Respiratory Viruses in Organ Cultures," *Brit. J. Exp. Pathol.* 46:109 (1965).
9. Ibrahim, H. L., Y. C. Zee and J. W. Osebold. "The Effects of Ozone on the Respiratory Epithelium and Alveolar Macrophage of Mice. I. Interferon Production," *Proc. Soc. Exp. Med.* 152:483 (1976).
10. Medin, N. I., J. W. Osebold and Y. C. Zee. "A Procedure for Pulmonary Lavage in Mice," *Am. J. Vet. Res.* 37:237 (1976).
11. Younger, J. S., and S. B. Salvin. "Production and Properties of Migration Inhibitory Factor and Interferon in the Circulation of Mice with Delayed Hypersensitivity," *J. Immunol.* 111:1914 (1973).
12. Valand, S. B., J. D. Acton and O. D. Myrvik. "Nitrogen Dioxide Inhibition of Viral-Induced Resistance in Alveolar Monocytes," *Arch. Environ. Health* 20:303 (1970).
13. Pattle, R. E., F. Burgess and H. Cullumbine. "The Effects of a Cold Environment and of Ammonia on the Toxicity of Sulfuric Acid Mist to Guinea Pigs," *J. Pathol. Bacteriol.* 72:219 (1956).

14. Amdur, M. O. "Aerosols Formed by Oxidation of Sulfur Dioxide," *Arch. Environ. Health* 23:459 (1971).

15. Amdur, M. O. "1974 Cummings Memorial Lecture. The Long Road from Donora," *J. Am. Ind. Hyg. Assoc.* 35:589 (1974).

16. Amdur, M. O. "The Impact of Air Pollutants on Physiologic Responses in the Respiratory Tract," *Proc. Am. Philos. Soc.* 114(1): 3 (1970).

17. Amdur, M. O. "The Respiratory Response of Guinea Pigs to Sulfuric Acid Mist," *Arch. Ind. Hyg. Occup. Med.* 18:407 (1958).

18. Amdur, M. O., L. Silverman and P. Drinker. "Inhalation of Sulfuric Acid Mist by Human Subjects," *Arch. Ind. Hyg. Occup. Med.* 6: 305 (1952).

19. Amdur, M. O., R. Z. Schulz and P. Drinker. "Toxicity of Sulfuric Acid Mist to Guinea Pigs," *Arch. Ind. Hyg. Occup. Med.* 5:318 (1952).

CHAPTER 11

EFFECTS OF CHRONIC EXPOSURE OF RATS TO AUTOMOBILE EXHAUST, H_2SO_4, SO_2, $Al_2(SO_4)_3$ AND CO

James P. Lewkowski and M. Malanchuk

U.S. Environmental Protection Agency
Laboratory Studies Division
Health Effects Research Laboratory
Cincinnati, Ohio 45268

Lloyd Hastings, A. Vinegar and G. P. Cooper

Department of Environmental Health
College of Medicine
University of Cincinnati
Cincinnati, Ohio 45267

Rats were exposed continuously for periods ranging from 6 to 14 weeks to catalytically treated automobile exhaust, untreated automobile exhaust, H_2SO_4, CO, SO_2 or $Al_2(SO_4)_3$. Automobile exhaust whether catalytically treated or not, significantly depressed the spontaneous locomotor activity (SLA) of rats, while exposure to H_2SO_4 and CO did not. Significant reductions in growth and in food and water consumption were observed only in rats exposed to noncatalytically treated automobile exhaust. Blood acid-base analysis showed that exposure to catalytically treated automobile exhaust or H_2SO_4 produced a metabolic alkalosis while exposure to CO caused a metabolic acidosis. Blood acid-base chemistry was within the normal range several weeks after termination of exposure. Independent studies were done in rats and guinea pigs exposed to H_2SO_4, SO_2 or $Al_2(SO_4)_3$. Pulmonary function studies indicated that the

$Al_2(SO_4)_3$ was most detrimental. Exposure to $Al_2(SO_4)_3$ having a mass median aerodynamic diameter (MMAD) of 2.0 μm resulted in increased static deflation volumes in juvenile rats, but exposure decreased the static deflation volume in the guinea pig. Exposure to either H_2SO_4 (MMAD of 0.5 μm or $Al_2(SO_4)_3$ caused an increased pulmonary resistance and respiration rate in adult rats. However, exposure to $Al_2(SO_4)_3$ having a MMAD of 1.4 μm caused primarily an increased respiratory rate and decreased lung compliance in adult rats. Spontaneous locomotor activity (SLA) during exposure and treadmill performance for up to 40 days postexposure was significantly decreased by exposure for $Al_2(SO_4)_3$ having a MMAD of 1.4 μm.

INTRODUCTION

Automobile emissions unquestionably reduce the quality of ambient air and, therefore, have been of concern to environmentalists for some time. The biological effects of exposure to automobile exhaust or to individual components of automobile exhaust such as carbon monoxide, nitrogen oxides, sulfur oxides, hydrocarbons and particulates, have been subjected to much experimental investigation and review.[1-9] In addition, various effects of chronic exposures to auto exhaust and other pollutant mixtures administered to beagles have been studied by Lewis et al.[10,11] and Vaughan et al.[12]

The outcome of this scrutiny has been the introduction of the catalytic converter, which is now required in most new automobiles. Catalytic converters have been shown to reduce pollutant emissions, which, in turn, reduce the biological effects.[13] However, health effects of automobile exhaust treated by catalytic converters is still of concern, particularly with respect to sulfuric acid (H_2SO_4) and related sulfur compounds. There is evidence that these compounds are toxic, as measured by their pulmonary effects. Acute exposure to approximately 20 ppm sulfur dioxide (SO_2) produces significant changes in pulmonary resistance in cats.[14] Amdur[15] has shown that H_2SO_4 is more irritating to the lungs than SO_2 and that the effect is dependent upon particle size. She also indicates that levels of SO_2 as low as 1 ppm have adverse effects on sensitive individuals, an effect that may be potentiated by the oxidation of SO_2 to H_2SO_4 by atmospheric particulates. Studies by Alarie et al.[16] indicate that changes in pulmonary function and structure in the cynomolgus monkey result from chronic exposure to 2.43 and 4.79 mg/m^3 H_2SO_4.

In these studies, we have measured and compared the health effects of emissions from catalyst-equipped engines, noncatalyst equipped engines, H_2SO_4, $Al_2(SO_4)_3$, CO and SO_2 on spontaneous locomotor activity, blood acid-base balance, food and water consumption, growth and pulmonary function.

EXPERIMENTAL PROCEDURES

Exposure Conditions-Exhaust Emissions Studies

The exposure system has been described previously.[17] In summary, a Ford engine was operated using indolene motor fuel that was supplemented with thiophene to provide a final sulfur concentration of 1000 ppm. The engine was run continuously (24 hr/day) during all three exposures, which included the 13-week and 6-week catalyst studies as well as a 6-week study without the catalyst. An Englehard/American Lava monolith catalyst was used in the exhaust system during the catalyst studies. Exhaust emissions were diluted with approximately 11 parts of clean air.

The aerometry data are summarized in Figures 1 and 2 and in Table I. It is evident that the catalytic converter substantially reduced the levels of carbon monoxide, hydrocarbons and nitrogen oxides in the exposure chambers. The catalyst was less effective in the later weeks of the 13-week catalyst study; the carbon monoxide levels rose and the sulfuric acid levels fell significantly. The acid concentration in the control sulfuric acid chambers was maintained at a high level. However, during the shorter 6-week catalyst study, the pollutant level did not change. The change in emissions in the former study must be kept in mind when interpreting the biological effects.

For the control sulfuric acid chambers, the H_2SO_4 was aerosolized using nebulizers and characterized by size and acidity. Concentration and size measurements were also made for the particulate matter in the auto exhaust atmosphere chambers. CO, nitrogen oxides (NO_x) and total hydrocarbons (HC) were automatically analyzed with online devices. In addition, the following exhaust components were measured periodically: methane, C_2-C_5 hydrocarbons, total aldehydes, SO_2, ammonia and ozone (O_3).

Exposure Conditions-Other Studies

In other studies (referred to hereafter as experiments A, B and C), three large (23 m^3) walk-in exposure chambers were used. Nominal airflow through each of the chambers was 2 m^3/min. One chamber was used to house control animals exposed to filtered air and the other two housed groups of experimental animals. In experiment A, rats were exposed to either H_2SO_4 mist, SO_2, or control air for 14 weeks. Experiments B and C were 7 and 8 weeks in duration, respectively. Both rats and guinea pigs were exposed to either H_2SO_4 mist or $Al_2(SO_4)_3$.

190

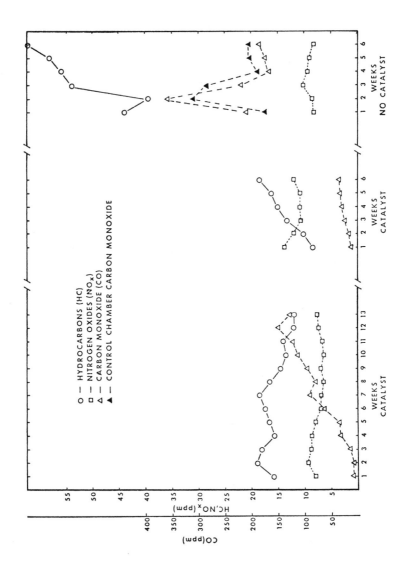

Figure 1. Mean weekly concentrations of total hydrocarbons (HC), nitrogen oxides (NO$_X$) and carbon monoxide (CO) during the three exhaust exposures.

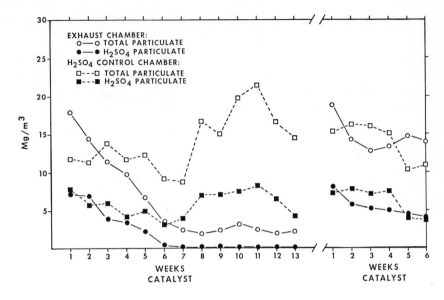

Figure 2. Mean weekly concentrations of particulates and H_2SO_4 in the exhaust and control chambers during the two catalyst-treated exhaust exposures.

In experiment A, SO_2 was obtained from compressed gas cylinders. Both H_2SO_4 and $Al_2(SO_4)_3$ particulates were obtained by nebulizing heated solutions[18] and injecting the mist into the main airflow. Concentrations were measured using various methodologies: SO_2 by the West-Gaeke technique,[19,20] total surfate by a turbidometric procedure and acid sulfate by titration. Particulate sulfate size was obtained by total sulfate or acid analyses of each of the several stages on an Anderson impactor.[21]

Procedure-Exhaust Emissions Studies

In all exhaust exposures, three groups of male weanling Sprague-Dawley rats were exposed to filtered air, automobile exhaust and a reference atmosphere (either H_2SO_4 mist or CO) at approximately the same level as in the exhaust atmosphere. Each of these three exposure conditions contained approximately 10 rats housed individually in standard 36-cm running wheels as well as approximatley 10 rats housed in standard cages. Some studies did not contain exactly 10 rats in both the wheel and cage groups in each exposure condition (*i.e.*, 7-10); the actual number depended on the constraints of the facility. However, there were equal numbers of rats in each exposure condition of a particular exhaust study.

Table I. Mean Concentrations of Atmospheric Components in the Exhaust Exposures

Pollutant	Atmosphere	13-Week Catalyst Study	6-Week Catalyst Study	6-Week Non-Catalyst Study
CO, ppm	Exhaust	70	25	210
	CO control	NA[a]	NA	230
THC, ppm as methane	Exhaust	16	14	50
NO, ppm	Exhaust	7.1	10.8	8.3
NO_2, ppm	Exhaust	0.3	0.4	5.1
GC-HCs, ppm Methane	Exhaust	3.23	4.07	3.85
Non-methane Aliphatics	Exhaust	1.56	1.33	4.41
Aldehydes, ppm	Exhaust	1.40	0.20	0.98
SO_2, ppm	Exhaust	2.22	1.31	2.42
Ammonia, ppm	Exhaust	0.044	0.033	0.016
Particulate, mg/m^3 Total	H_2SO_4 control	14.85	15.18	0.0 (CO control)
	Exhaust	5.65	13.65	0.93
Sulfate	H_2SO_4 control	8.80	7.05	0.0 (CO control)
	Exhaust	3.06	5.91	0.14
Sulfuric Acid	H_2SO_4 control	6.59	6.35	0.0
	Exhaust	1.74	4.93	0.0
Particulate Size μm, MMD	H_2SO_4 control	0.31	0.44	NA
	Exhaust	0.24	0.25	0.17

[a]NA = not applicable.

The number of wheel revolutions was recorded daily. Food and water intake and body weight of both the animals in the activity wheels and in the cages were reported weekly. Exposure was continuous except for the time required for the cleaning and changing of food and water (approximately six hours per week). Near the end of the exposure period, blood samples were obtained from the orbital sinuses of the rats housed in the standard cages and were used to determine the acid-base status. The blood analysis (London Radiometer) included pH, pCO_2, hematocrit and estimation of HCO_3^- and buffer base.

All biological parameters were measured after a recovery period in which all groups were exposed to filtered air. The analysis of variance test was used to analyze all data except the acid-base data, which **were** analyzed using the t test.

Procedures-Studies A, B and C

General

In these experiments, weekly measurements of food and water consumption and body weight were taken as well as were daily measurements of spontaneous locomotor activity, as described above. At the end of the exposure period, blood chemistry tests were performed in experiments A and B as described above. In addition, studies of learning ability were done in experiments A and B, and pulmonary function tests were performed at the termination of the exposure of all of these studies. Further details of the procedures used for assessing learning ability are presented in the Results section.

Pulmonary Function

The following techniques are modifications of ones described by Diamond.[22-24] Animals were anesthetized with urethane [1 g/kg body weight intraperitoneal (IP)] given in a 25% solution. One arm of an L-shaped stainless steel cannula was placed in the trachea and the other arm of the cannula was connected to a Fleisch pneumotachograph. A differential pressure transducer measured the pressure drop across the pneumotachograph screen. The pressure drop was calibrated as a flow, and the flow signal was integrated to obtain tidal volume. A water-filled cannula was inserted in the esophagus with the open end of the cannula located just anterior to the stomach. The other end of the cannula was connected to one side of a differential pressure transducer outside the plethysmograph and the other side of the transducer was connected to the tracheal cannula. Thus, the pressure recorded was representative of the transpulmonary pressure. Dynamic compliance was determined from values of volume and pressure taken at points of zero flow. Pulmonary resistance was determined from values of pressure and flow taken at equal mid-tidal volume points.

After the dynamic measurements were made, the animal was sacrificed with an overdose of urethane and the lungs were exposed by cutting through the sternum and retracting the ribs. The tracheal cannula was then connected to a water manometer and a syringe. The lungs were inflated to 30 cm H_2O air pressure and deflated to atmospheric pressure. This procedure was done twice to give the lungs a volume history. They were then inflated and maintained at 30 cm H_2O for 5 seconds. Then air was withdrawn to reduce the pressure to 25 cm H_2O for 15 seconds and the volume was recorded. This process was repeated at 5-cm H_2O increments down to atmospheric pressure.

RESULTS

Exhaust Emissions Studies

Spontaneous Locomotor Activity

It is evident that exposure to the exhaust atmosphere, whether catalytically treated or not, produced a marked decrease in spontaneous activity when compared to the controls (Figure 3). The H_2SO_4 exposure atmosphere had no consistent effect on activity. In addition, there was a rebound in the activity after the exhaust exposure was terminated in the noncatalyst study. It is also interesting that the CO exposure alone in the noncatalyst study caused an increase in the spontaneous activity. Thus one may postulate that this CO level in the exhaust atmosphere may have counteracted the depressant effects of other components in the exhaust.

Food and Water Consumption, Body Weight

There were no significant differences from control in food and water consumption or body weight in either of the catalyst exhaust exposures (Figure 4). However, exposure to noncatalyst exhaust resulted in a significant reduction in food and water consumption as well as a reduction in body weight. These changes were not due to CO exposure since they did not occur in the group exposed to CO alone.

Blood Chemistry

Exposure to either exhaust or sulfuric acid in both the 6 and 13-week catalyst studies elicited some acid-base changes typical of a metabolic alkalosis (Table II). Exposure to exhaust from the noncatalyst study had less of an effect; however, it should be noted that exposure to CO alone caused a shift in the acid-base pattern toward metabolic acidosis. Thus, the effects of exhaust alone on acid-base balance may be neutralized by a high level of CO in the same atmosphere. In all studies, the acid-base parameters were within the normal range one to three weeks after termination of exposure.

Experiment A: Fourteen-Week Exposure to H_2SO_4 and SO_2

Aerometry

The aerometry data for this experiment are summarized in Table III. The total sulfate, acid sulfate, H_2SO_4 particle size and SO_2 concentrations

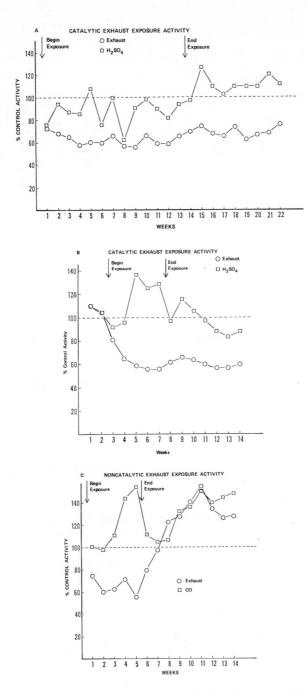

Figure 3. Spontaneous locomotor activity (SLA) during the three exhaust exposures: A, 13-week catalyst study; B, 6-week catalyst study; C, 6-week non-catalyst study.

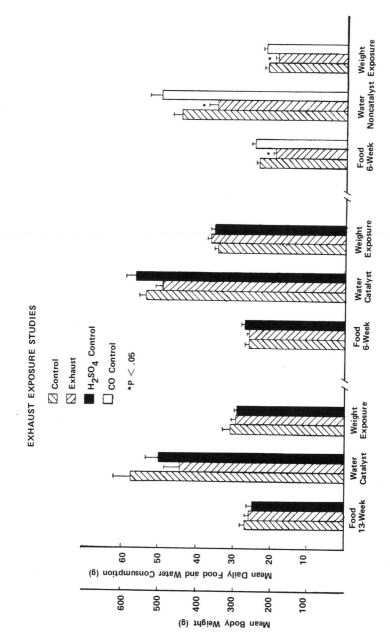

Figure 4. Mean food and water consumption and body weight data from the three exhaust exposures.

Table II. Blood Chemistry of Rats Exposed in the Exhaust Studies. Values are Mean ± Standard Error. Base Excess (BE) and HCO_3^- are in meq/l. The Number of Rats per Group is Shown in Parentheses

A. 13-Week Catalyst Study

Exposure	pH	pCO_2	BE	HCO_3^-
Control	7.35 ± 0.04 (6)	47.0 ± 2.47 (10)	0.2 ± 1.73 (9)	26.2 ± 1.68 (9)
H_2SO_4	7.47 ± 0.003[a] (7)	51.9 ± 2.17 (9)	11.4 ± 0.97[a] (9)	38.3 ± 1.48[a] (9)
Exhaust	7.46 ± 0.01[a]	54.3 ± 1.10[b] (9)	12.0 ± 0.99[a] (9)	39.1 ± 1.16[a] (9)

B. 6-Week Catalyst Study

Exposure	pH	pCO_2	BE	HCO_3^-
Control	7.36 ± 0.02 (9)	52.1 ± 2.26 (9)	2.4 ± 1.26 (9)	29.1 ± 1.36 (9)
H_2SO_4	7.34 ± 0.02 (10)	49.2 ± 1.55[b] (9)	3.8 ± 1.32 (10)	32.3 ± 1.44 (10)
Exhaust	7.41 ± 0.01[b] (10)	52.6 ± 2.22 (9)	6.9 ± 1.32[b] (10)	33.5 ± 1.59 (10)

C. 6-Week Noncatalyst Study

Exposure	pH	pCO_2	BE	HCO_3^-	Hematocrit
Control	7.38 ± 0.01 (10)	50.4± 2.49 (8)	4.1 ± 1.32 (8)	30.5 ± 1.61 (8)	49.3 ± 1.64 (7)
Carbon Monoxide	7.32 ± 0.02 (10)	45.9 ± 2.30 (10)	-3.2 ± 0.98[a] (10)	23.4 ± 0.75[a] (10)	56.7 ± 0.93[a] (9)
Exhaust	7.35 ± 0.02 (10)	50.7 ± 1.30 (10)	1.1 ± 1.15 (10)	28.0 ± 1.10 (10)	52.9 ± 0.81 (9)

[a]$P < 0.01$.
[b]$P < 0.05$.

Table III. Aerometry Data for the 14-Week Exposure to H_2SO_4 and SO_2.
Values are Mean ± Standard Error

	Control Chamber	H_2SO_4 Chamber	SO_2 Chamber
Total Sulfate	–	2.37 ± 1.13 mg/m^3	–
Particle Size, MMD	–	0.5 ± 1.0 μm	–
SO_2	–	–	0.89 ± 0.39 mg/m^3
Relative Humidity	42.92 ± 5.62	49.81 ± 5.26	50.8 ± 5.61
Temperature	$78.23 \pm 1.97°F$	$76.92 \pm 2.14°F$	$76.16 \pm 3.22°F$

were sampled three times a week while the relative humidity and temperature were measured daily.

Growth, Activity and Learning

In this experiment, 30 male Sprague-Dawley rats were randomly divided into three equal groups. The rats were placed in standard 36-cm running wheels and exposed to either control air, H_2SO_4 or SO_2. During exposure, food and water intake, as well as body weight, were recorded weekly. The number of wheel revolutions was recorded daily.

At the end of the exposure, the rats were placed on water deprivation until they dropped to 85% of baseline body weight. The rats were then placed in operant chambers and shaped to bar press for liquid reinforcement. Once that response was learned, the rats were tested on a discrete trial successive brightness discrimination task in Experiment A. Briefly, in this task, when the houselight came on in the bright mode, the rat had to press the left bar; when it came on in the dim mode, pressing the right bar was reinforced. The rats received 25 trials per day, 5 days a week and were allowed to correct their mistakes. All programming was controlled by Coulbourn logic modules.

Analysis of variance revealed no significant difference among the three groups in amount of food or water consumed, body weight or activity levels, as shown in Figure 5 and 6. Thus, exposure to either SO_2 or H_2SO_4 at the levels used in this study appears to have little if any effect on these parameters. Figure 7 depicts the results of the acquisition of the discrimination task by three groups of rats. Again, no differences were evident among the three groups. However, it should be noted that the task was apparently so difficult that little was learned by any group, even after a month of testing. For this reason, a more simple task was selected for use in experiment B.

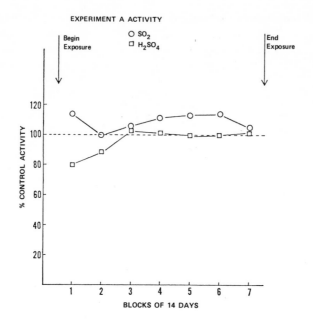

Figure 5. Spontaneous locomotor activity (SLA) during the 14-week exposure to H_2SO_4 and SO_2.

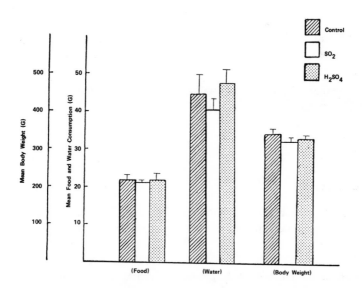

Figure 6. Mean daily food and water consumption and body weight during the 14-week exposure to SO_2 and H_2SO_4.

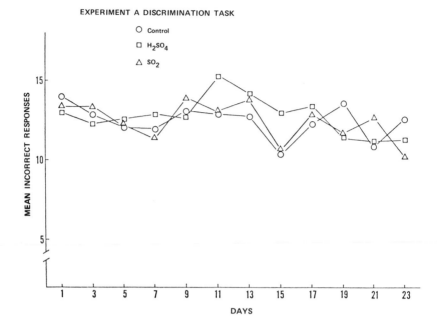

Figure 7. Effects of H_2SO_4 and SO_2 exposure on the acquisition of a discrimination task.

Pulmonary Function

As shown in Table IV, no significant differences were found between groups of juvenile rats for any of the pulmonary function parameters. Tidal volumes of the SO_2-exposed adult rats were significantly higher than the control animals ($p < 0.01$).

Blood Chemistry

Table V summarizes the blood gas data. No differences in blood gas parameters were detected as a result of exposure to either SO_2 or H_2SO_4.

Experiment B: Seven-Week Exposure to H_2SO_4 and $Al_2(SO_4)_3$

Table IV. Dynamic Pulmonary Function Parameters of Rats Exposed in the 14-Week Exposure to H_2SO_4 and SO_2. Values Indicated are Mean ± Standard Error or Percent of Control

	Weight (g)	Tidal Volume (ml)	Respiratory Frequency (min⁻¹)	Resistance (% Control)	Resistance x Weight (% Control)	Compliance (% Control)	Compliance/ Weight (% Control)
Juvenile Rats							
Control	175 ± 6	1.69 ± 0.06	170.5 ± 8.6	100	100	100	100
SO_2	179 ± 5	1.76 ± 0.05	159.8 ± 9.7	83	86	94	92
H_2SO_4	189 ± 7	1.77 ± 0.07	166.5 ± 7.3	95	104	102	98
Adult Rats							
Control	390 ± 8	2.57 ± 0.09	140.3 ± 4.7	100	100	100	100
SO_2	413 ± 11	2.98 ± 0.07[a]	129.9 ± 7.0	84	93	103	98
H_2SO_4	425 ± 11[b]	2.71 ± 0.06	142.5 ± 3.3	102	113	108	99

[a] $p < 0.01$.
[b] $p < 0.05$.

Table V. Blood Chemistry of Rats Exposed in the 14-Week H_2SO_4 and SO_2 Study. Values are Mean ± Standard Error. Base Excess (BE) and HCO_3^- are in meq/l

Exposure	pH	pCO_2	Hematocrit	BE	HCO_3^-
Control	7.27 ± 0.03	41 ± 1.9	54 ± 1.5	-8.3 ± 1.9	18.5 ± 1.5
SO_2	7.32 ± 0.02	40 ± 1.7	52 ± 0.9	-5.6 ± 1.1	19.9 ± 1.0
H_2SO_4	7.29 ± 0.02	45 ± 2.4	52 ± 1.6	-5.3 ± 0.9	21.4 ± 0.8

Table VI. Aerometry Data for the Seven-Week Exposure to H_2SO_4 and $Al_2(SO_4)_3$. Values are Mean ± Standard Error

	Control Chamber	H_2SO_4 Chamber	$Al_2(SO_4)_3$ Chamber
Total Sulfate	–	4.05 ± 1.63 mg/m³	–
Aluminum Sulfate	–	–	2.04 ± 1.07 mg/m³
Particle Size, MMD	–	0.5 ± 1.0 μm	2.0 ± 1.97 μm
Relative Humidity	27.36 ± 7.2	33.3 ± 6.78	36.95 ± 8.77
Temperature	72.52 ± 1.08°F	70.75 ± 1.96°F	69.2 ± 2.21°F

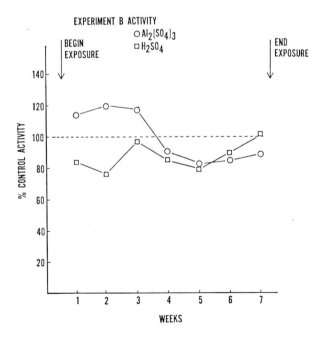

Figure 8. Spontaneous locomotor activity during the seven-week exposure to H_2SO_4 and $Al_2(SO_4)_3$.

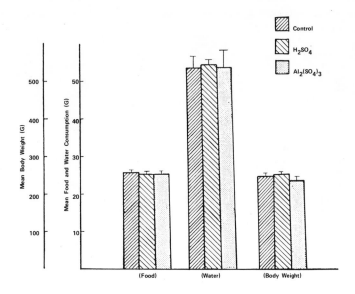

Figure 9. Mean daily food and water consumption and body weight during the seven-week exposure to H_2SO_4 and $Al_2(SO_4)_3$.

Aerometry

The particle size and H_2SO_4 and $Al_2(SO_4)_3$ concentrations were measured three times a week and the relative humidity and temperature were measured daily. The data are summarized in Table VI.

Growth, Activity and Learning

Thirty-six male Sprague-Dawley rats were used in this experiment. Three groups of twelve rats each were placed into the running wheels and exposed to either control air, H_2SO_4 or $Al_2(SO_4)_3$. Again, food and water intake, as well as body weight, were recorded weekly and wheel revolutions were recorded daily.

After termination of exposure, the rats were water deprived and taught to bar press for liquid reinforcement. They were then tested for the acquisition of a cued spatial discrimination task. In this task, two bars were extended into the chamber continuously. At the onset of the house-light, pressing the right bar yielded reinforcement. However, pressing the left scored an incorrect response. The animals received 50 trials per day. After six days of acquisition training, a reversal procedure was initiated, that is, pressing the left bar at houselight onset produced reinforcement.

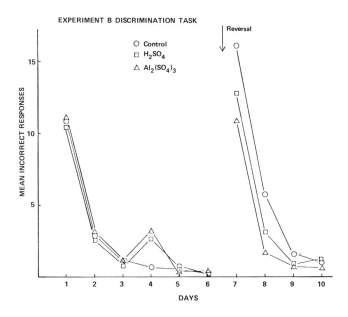

Figure 10. Effects of H_2SO_4 and $Al_2(SO_4)_3$ exposure on the acquisition of a spatial discrimination task.

As in the previous experiment, analysis of the data revealed no significant differences among the groups in amount of food and water consumed, body weight or activity levels as shown in Figures 8 and 9.

None of the parameters examined appeared to be affected in any way by the different exposure conditions. The same was true for the acquisition of the spatial discrimination task (Figure 10). By day 6, all three groups were performing the task almost perfectly. On day 7 the reversal procedure was initiated. Although there was a slight but nonsignificant difference at the beginning of the reversal, performance of the groups quickly stabilized and all had mastered the task by day 4. A second reversal was acquired almost equally by the three groups in less than 75 trials. Exposure to the various compounds did not affect any other parameters, nor was there any evidence of change in the higher cortical functions, such as learning, as measured by these tasks.

Pulmonary Function

The measurements of dynamic pulmonary function parameters are summarized in Table VII and the static deflation volumes are shown in Figures 11-13.

Table VII. Dynamic Pulmonary Function Parameters of Rats Exposed in the Seven-Week Exposure to H_2SO_4 and $Al_2(SO_4)_3$. Values Indicated are Mean ± Standard Error or Percent of Control

	Weight (g)	Tidal Volume (ml)	Respiratory Frequency (min^{-1})	Resistance (% Control)	Resistance x Weight (% Control)	Compliance (% Control)	Compliance/Weight (% Control)
Juvenile Rats							
Control	207 ± 7	1.98 ± 0.08	144.3 ± 8.9	100	100	100	100
H_2SO_4	197 ± 10	1.81 ± 0.09	146.2 ± 9.9	170	159	98	110
$Al_2(SO_4)_3$	210 ± 7	1.81 ± 0.06	174.2 ± 14.9	158	161	156	153
Adult Rats							
Control	391 ± 9	2.65 ± 0.05	130 ± 6	100	100	100	100
H_2SO_4	383 ± 11	2.28 ± 0.10[a]	153 ± 5[a]	192	189[b]	110	115
$Al_2(SO_4)_3$	361 ± 19	2.34 ± 0.19	159 ± 8[c]	188	172[c]	121	130
Guinea Pigs							
Control	543 ± 22	3.00 ± 0.66	50 ± 4	100	100	100	100
H_2SO_4	472 ± 21[d]	2.81 ± 0.10	56 ± 4	86	72	96	108
$Al_2(SO_4)_3$	497 ± 30	3.03 ± 0.15	59 ± 6	240	230	85	91

[a] $p < 0.01$.
[b] $p < 0.001$.
[c] $p < 0.02$.
[d] $p < 0.05$.

Table VIII. Blood Chemistry of Rats Exposed in the Seven-Week H_2SO_4
and $Al_2(SO_4)_3$ Study. Values are Mean $\pm$ Standard Error.
Base Excess (BE) and HCO_3^- are in meq/l

Exposure	pH	pCO_2	Hematocrit	BE	HCO_3^-
Control	7.42 ± 0.01	46 ± 1.4	57 ± 1.2	4.2 ± 0.9	29.9 ± 1.1
H_2SO_4	7.42 ± 0.01	45 ± 1.5	57 ± 1.2	4.3 ± 0.7	29.9 ± 1.0
$Al_2(SO_4)_3$	7.41 ± 0.01	47 ± 1.1	58 ± 0.9	4.2 ± 0.9	30.2 ± 0.9

The pulmonary resistance of adult rats exposed to H_2SO_4 and
$Al_2(SO_4)_3$ was significantly higher than controls ($p < 0.001$ and $p <$
0.01, respectively). Tidal volumes of H_2SO_4 animals were significantly
lower than controls ($p < 0.01$). Respiratory frequencies of H_2SO_4 and
$Al_2(SO_4)_3$ animals were significantly higher than controls ($p < 0.01$ and
$p < 0.02$, respectively). Guinea pig pulmonary resistance was somewhat
higher but this increase did not meet the 0.05 criterion of statistical
significance. Static deflation volumes were significantly lower than controls in $Al_2(SO_4)_3$ animals ($p < 0.05$). In juvenile rats exposed to
H_2SO_4 and $Al_2(SO_4)_3$, there also occurred increases in pulmonary
resistance, respiratory frequency, functional residual capacity and compliance and a decrease in tidal volume, although these changes did not
quite meet the 0.05 level of significance. Static lung deflation volumes
were significantly higher in rats exposed to $Al_2(SO_4)_3$.

Blood Chemistry

Table VIII summarizes the data from rats exposed to control air
H_2SO_4 and $Al_2(SO_4)_3$ in this study. There were not significant differences
between groups for any of the acid-base parameters.

Experiment C: Eight-Week Exposure to H_2SO_4 and $Al_2(SO_4)_3$

Aerometry

The exposure conditions are summarized in Table IX. The main differences between experiments B and C were the smaller particle sizes of
H_2SO_4 and $Al_2(SO_4)_3$ and the slightly higher concentration of $Al_2(SO_4)_3$
in experiment C.

Growth and Activity

Analysis of the food and water consumption of the adult rats showed

Table IX. Aerometry Data for the Eight-Week Exposure to H_2SO_4 and $Al_2(SO_4)_3$. Values are Mean ± Standard Error

	Control Chamber	H_2SO_4 Chamber	$Al_2(SO_4)_3$ Chamber
Total Sulfate	–	2.49 mg/m^3 ± 0.4	–
Aluminum Sulfate	–	–	2.59 mg/m^3 ± 1.4
Particle Size, MMD	–	< 0.24 μm	1.4 μm ± 2.3
Relative Humidity	49.8 ± 0.70	43.8 ± 0.90	56.4 ± 1.06
Temperature	71.0 ± 0.12	73.6 ± 0.17	65.3 ± 0.59

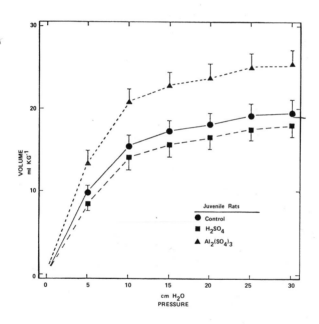

Figure 11. Static deflation volumes of juvenile rats exposed to H_2SO_4 and $Al_2(SO_4)_3$.

Experiment C: Eight-Week Exposure to H_2SO_4 and $Al_2(SO_4)_3$

that the $Al_2(SO_4)_3$ group consumed significantly more than the controls ($p < 0.05$). Body weight gain was equal for all groups throughout the exposure (Figure 14).

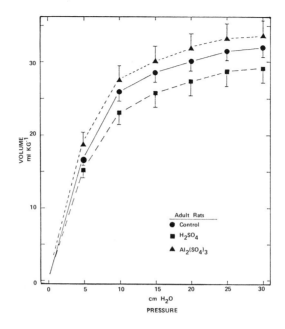

Figure 12. Static deflation volumes of adult rats exposed to H_2SO_4 and $Al_2(SO_4)_3$.

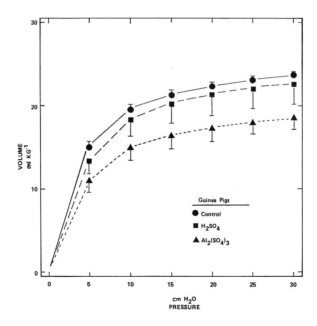

Figure 13. Static deflation volumes of guinea pigs exposed to H_2SO_4 and $Al_2(SO_4)_3$.

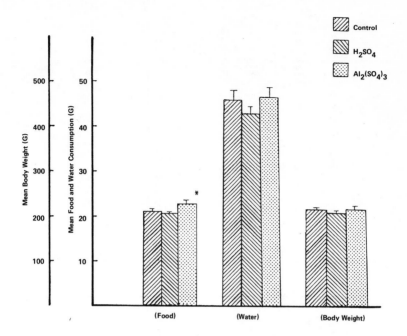

Figure 14. Mean daily food and water consumption and body weight during the eight-week exposure to H_2SO_4 and $Al_2(SO_4)_3$.

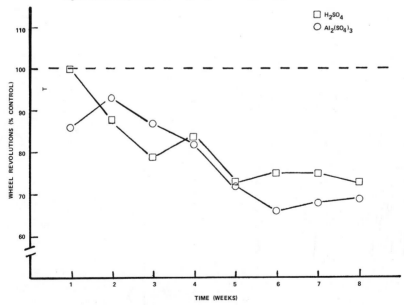

Figure 15. Spontaneous locomotor activity during the eight-week exposure to H_2SO_4 and $Al_2(SO_4)_3$.

Table X. Performance on a Forced Activity Task (Treadmill) Immediately on Termination of the Eight-Week Exposure to H_2SO_4 and $Al_2(SO_4)_3$ and 40 and 80 Days Postexposure

	At Termination of Exposure	40 Days Post-exposure	80 Days Post-exposure
A. Mean number of seconds (± SEM) to exhaustion on forced activity task			
Control	4704 ± 662	2275 ± 229	1685 ± 248
H_2SO_4	3365 ± 492	1906 ± 344	1448 ± 194
$Al_2(SO_4)_3$	2761 ± 395[a]	1448 ± 194[a]	1364 ± 295
B. Mean number of shocks (± SEM) received during forced activity task			
Control	206 ± 40	273 ± 51	281 ± 63
H_2SO_4	214 ± 53	471 ± 82	436 ± 116
$Al_2(SO_4)_3$	303 ± 59	547 ± 81[a]	507 ± 50[a]

[a] $p < 0.05$

Table XI. Dynamic Pulmonary Function Parameters of Rats Exposed in the Eight-Week Exposure to H_2SO_4 and $Al_2(SO_4)_3$. Values Indicated are Mean ± Standard Error or Percent of Control

	Weight (g)	Tidal Volume (ml)	Respiratory Frequency (min^{-1})	Resistance (% Control)	Resistance x Weight (% Control)	Compliance (% Control)	Compliance/ Weight (% Control)
Juvenile Rats							
Control	259 ± 4	2.13 ± 0.08	134 ± 7	100	100	100	100
H_2SO_4	253 ± 2	2.19 ± 0.07	136 ± 5	92	89	94	97
$Al_2(SO_4)_3$	255 ± 9	2.11 ± 0.04	154 ± 8	78	76	105	106
Adult Rats							
Control	362 ± 6	2.30 ± 0.05	143 ± 8	100	100	100	100
H_2SO_4	358 ± 6	2.47 ± 0.06[a]	132 ± 5	101	102	92	93
$Al_2(SO_4)_3$	349 ± 6	2.20 ± 0.09	172 ± 7[a]	97	94	70[b]	72[a]
Guinea Pigs							
Control	705 ± 15	2.60 ± 0.19	117 ± 8	100	100	100	100
H_2SO_4	700 ± 19	2.45 ± 0.19	114 ± 9	94	93	105	106
$Al_2(SO_4)_3$	704 ± 20	2.58 ± 0.18	122 ± 10	83	83	113	112

[a] $p < 0.05$.
[b] $p < 0.01$.

Analysis of the spontaneous locomotor activity using analysis of variance indicated a significant interaction effect ($p < 0.05$). Inspection of the data (Figure 15) shows that the controls ran at a faster rate than did rats exposed to either H_2SO_4 or $Al_2(SO_4)_3$.

In addition to the measurements of spontaneous locomotor activity made during the period of exposure, the rats were also tested in this experiment on a forced activity task using a motorized treadmill. Tests were conducted at the termination of exposure and at 40 and 80 days postexposure. As shown in Table X, major differences were observed in performance between the control and $Al_2(SO_4)_3$ groups. Those rats exposed to $Al_2(SO_4)_3$ reached the point of exhaustion in a significantly shorter time ($p < 0.05$) and also received nearly 50% more shocks while on the treadmill for a shorter time period. This difference in time to exhaustion decreased during the postexposure tests but the number of shocks to the $Al_2(SO_4)_3$ exposed group was increased ($p < 0.05$).

Pulmonary Function

Dynamic pulmonary function test results are shown in Table XI. Except for the increased respiratory frequency ($p < 0.05$) and decreased lung compliance ($p < 0.01$) in the $Al_2(SO_4)_3$ group of adult rats, there appears to be little change.

Blood Chemistry

Table XII summarizes the blood acid-base analyses at the termination of experiment C. Except for a slight, statistically insignificant, tendency toward alkalosis in animals exposed to $Al_2(SO_4)_3$, there were no differences between groups.

Table XII. Blood Chemistry of Rats Exposed in the Eight-Week H_2SO_4 and $Al_2(SO_4)_3$ Study. Values are Mean ± Standard Error. Base Excess (BE) and HCO_3^- are in meq/l

Exposure	pH	pCO_2	Hematocrit	BE	HCO_3^-
Control	7.25 ± 0.09	47.00 ± 5.43	54.11 ± 1.61	-6.65 ± 5.18	20.80 ± 3.60
H_2SO_4	7.32 ± 0.06	43.50 ± 3.59	53.22 ± 1.09	-3.65 ± 3.65	22.40 ± 2.70
$Al_2(SO_4)_3$	7.36 ± 0.05	44.11 ± 5.53	51.00 ± 2.36	0.93 ± 3.20	24.55 ± 2.95

DISCUSSION AND CONCLUSIONS

Automobile exhaust exposure results in a depression of spontaneous locomotor activity of rats while exposure to atmospheres in the positive control chambers (either H_2SO_4 or CO) does not depress the spontaneous activity. In fact, CO increased the activity to some extent. Similar results have been noted previously.[26,27] Consistent results were found in other exposure studies. Thus, in both experiment A (H_2SO_4 and SO_2) and experiment B (H_2SO_4 and $Al_2(SO_4)_3$, there were no differences observed in either spontaneous locomotor activity or learning ability. In addition, the animals exposed to CO, SO_2, H_2SO_4 and $Al_2(SO_4)_3$ did not exhibit any statistically significant differences in food and water consumption or body weight. However, in experiment C, there were reductions in spontaneous locomotor activity in animals exposed to $Al_2(SO_4)_3$ and H_2SO_4. Furthermore, immediately following termination of exposure, the ability of the $Al_2(SO_4)_3$ animals to perform maximally on a treadmill was also greatly impaired. The smaller particle sizes in this study compared with previous experiments might have brought about such change.

One must conclude that components other than CO, SO_2 or H_2SO_4 are primarily responsible for the depressant effects of automobile exhaust on activity. Others have suggested that the hydrocarbon levels in automobile exhaust might be an important causative factor.[28] However, the action of other exhaust components or an interaction of CO, SO_2 or H_2SO_4 with other exhaust components to produce the reduction in activity cannot be excluded.

Blood acid-base analyses showed a shift toward a metabolic alkalosis in the animals exposed to the catalytically treated automobile exhaust or H_2SO_4 in both the 13- and 6-week studies. The alkaline shift may have been the result of excess acid excretion by the kidneys. Others have observed an increased urine acidity in workers exposed to relatively high levels of SO_2.[29] Acid sulfate alone cannot be the causative factor since there was virtually no H_2SO_4 present in the exposure chamber during the last 6 weeks of the 13-week catalyst study. However, in experiments A, B and C, no differences in blood chemistry were seen between control rats and rats exposed to SO_2, H_2SO_4 or $Al_2(SO_4)_3$, possibly because of the lower concentrations used in these studies.

The acid-base pattern of animals exposed to the noncatalyst exhaust showed fewer effects. However, the animals exposed to CO alone demonstrated a shift in blood chemistry toward metabolic acidosis. A similar effect of CO has been observed previously.[30] The lack of an effect in the exhaust chamber may be a result of the effects of CO and other exhaust components such as SO_2 cancelling, thus causing little net change. It is evident that the etiology of the acid-base changes may be determined only through further exposure to other exhaust components such as hydrocarbons, nitrogen oxides and various sulfur components, and through analysis of concomitant renal function studies.

The results of experiment B indicate that both H_2SO_4 and $Al_2(SO_4)_3$ have an effect on pulmonary function. Pulmonary resistance and respiratory frequencies of adult rats were increased by both agents. However, $Al_2(SO_4)_3$ not only increased respiratory frequencies of juvenile rats but also increased the functional residual capacity and static deflation volumes. In experiment C an increase in respiratory frequency and a decrease in lung compliance from $Al_2(SO_4)_3$ was observed.

Increased static deflation volumes shown by the lungs of juvenile rats exposed to aluminum sulfate in experiment B can be attributed to a loss of elastic recoil or gas trapping. The functional residual capacity was higher in the aluminum sulfate-exposed animals, but not significantly. Thus, the reason for the higher static deflation volumes cannot be specified precisely. The decreased static deflation volumes of the lungs of guinea pigs exposed to $Al_2(SO_4)_3$ is attributable to the diffuse hemorrhagic areas seen during gross examination of the lungs. These results lead to the conclusion that $Al_2(SO_4)_3$ has a more detrimental effect on the lung than does H_2SO_4 in both juvenile and adult rats.

A comparison of the results of the catalyst and noncatalyst exposure clearly indicates the reduction in biological effects with the catalytic converter. Only the noncatalyst exhaust exposure resulted in significant changes in activity, body weight and food and water consumption. Such effects were not observed in exposure to sulfuric acid even though the levels of exposure were of comparable magnitude. From experiments A, B and C, it appears that the aluminum cation is more important in determining toxicity than either the hydrogen ion or sulfate anion. Most of the effects observed, including the depression of activity, can possibly be a result of insults to the lung. Further work will be required to substantiate adequately these conclusions.

ACKNOWLEDGMENTS

We thank Dr. L. Kleinman for advice and assistance in the blood gas analysis; Dr. S. Brooks for advice and assistance with the pulmonary studies; T. Disney, J. Williams, T. Wessendarp, J. Mann, L. Galea and

R. Sadis for technical assistance; and D. Dean, V. Tilford and J. Roe for clerical assistance. This work was supported in part by EPA Contracts 68-03-0492 and 68-03-2335 and NIEHS grant ES00159.

REFERENCES

1. Murphy, S. D. "A Review of Effects on Animals of Exposure to Auto Exhaust and Some of the Components," *J. Air Poll. Control Assoc.* 14:303-307 (1964).
2. Hueter, F. G., G. L. Contner, K. A. Busch and R. G. Hinners. "Biological Effects of Atmospheres Contaminated by Auto Exhaust," *Arch. Environ. Health* 12:553-560 (1966).
3. Emik, L. A., and R. L. Plata. "Depression of Running Activity in Mice by Exposure to Polluted Air," *Arch. Environ. Health* 13:574-579 (1969).
4. Lewis, T. R., F. G. Hueter and K. A. Busch. "Irradiated Automotive Exhaust. It's Effects on the Reproduction of Mice," *Arch. Environ. Health* 15:26-35 (1967).
5. U.S. Department of Health, Education and Welfare, Public Health Service, National Air Pollution Control Administration. "Air Quality Criteria for Carbon Monoxide." (Washington, DC: U.S. Government Printing Office, 1970).
6. Environmental Protection Agency, Air Pollution Control Office. "Air Quality Criteria for Nitrogen Oxides." (Washington, DC: U.S. Government Printing Office, 1971).
7. U.S. Department of Health, Education and Welfare. "Air Quality Criteria for Sulfur Oxides," Public Health Service, National Air Pollution Control Administration. (Washington, DC: U.S. Government Printing Office, 1964).
8. U.S. Department of Health, Education and Welfare. "Air Quality Criteria for Hydrocarbons." Public Health Service, National Air Pollution Control Administration. (Washington, DC: U.S. Government Printing Office, 1970).
9. U.S. Department of Health, Education and Welfare. "Air Quality Criteria for Particulate Matter," Public Health Service, National Air Pollution Control Administration. (Washington, DC: U.S. Government Printing Office, 1969).
10. Lewis, T. R., W. J. Moorman, W. F. Ludmann and K. I. Campbell. "Toxicity of Long-Term Exposure to Oxides of Sulfur," *Arch. Environ. Health* 26:16-23 (1973).
11. Lewis, T. R., W. J. Moorman, Y. Y. Yang and J. F. Stara. "Long-Term Exposure to Auto Exhaust and Other Pollutant Mixtures," *Arch. Environ. Health* 29:102-106 (1974).
12. Vaughan, T. R., L. F. Jennelle and T. R. Lewis. "Long-Term Exposure to Low Levels of Air Pollutants," *Arch. Environ. Health* 19:45-50 (1969).

13. Hysell, D. K., W. Moore, R. Hinners, M. Malanchuk, R. Miller and J. F. Stara. "Inhalation Toxicology of Automotive Emissions as Affected by an Oxidation Exhaust Catalyst," *Environ. Health Pers.* 10:57-62 (1975).

14. Corn, M., M. Kotsko, D. Stanton, W. Bell and A. P. Thomas. "Response of Cats to Inhaled Mixtures of SO_2-NaCl Aerosol in Air," *Arch. Environ. Health* 24:248-256 (1972).

15. Amdur, M. O. "Toxicological Appraisal of Particulate Matter Oxides of Sulfur, and Sulfuric Acid," *J. Air Poll. Control Assoc.* 19:638-644 (1975).

16. Alarie, Y., W. M. Busey, A. A. Krumm and C. E. Ulrich. "Long-Term Continuous Exposure to Sulfuric Acid Mist in Cynomolgus Monkeys and Guinea Pigs," *Arch. Environ. Health* 27:16-24 (1973).

17. Hinners, R. G., J. K. Burkart and C. L. Punte. "Animal Inhalation Exposure Chambers," *Arch. Environ. Health* 16:194-206 (1968).

18. American Conference of Governmental Industrial Hygienists, Committee on Air Sampling Instruments. *Air Sampling Instruments for Evaluation of Atmospheric Contaminants,* 4th ed., Cincinnati, OH (1972), pp. 114-120.

19. West, P. W., and G. C. Gaeke. "Fixation of Sulfuric Dioxide as Sulfitomercurate III and Subsequent Colorimetric Determination," *Anal. Chem.* 28:1816 (1956).

20. Intersociety Committee. *Methods of Air Sampling and Analysis* (Washington, DC: American Public Health Association, Inc., 1972), pp. 447-455.

21. Flesch, J. P., C. H. Norris and A. E. Nugent, Jr. "Calibrating Particulate Air Samples with Monodispense Aerosols: Application to the Anderson Cascade Impactor," *Am. Ind. Hyg. Assoc. J.* 28:507-516 (1967).

22. Diamond, L. "Utilization of Changes in Pulmonary Resistance for the Evaluation of Bronchodilator Drugs," *Arch. Int. Pharmacodyn.* 168:239-250 (1967).

23. Diamond, L. "Potentiation of Bronchomotor Responses by Beta Adrenergic Antagonists," *J. Pharmacol. Exp. Ther.* 181:434-445 (1972).

24. Diamond, L., G. K. Adams, B. Bleidt and B. Williams. "Experimental Study of a Potential Anti-Asthmatic Agent: SCH 15280," *J. Pharmacol. Exp. Ther.* 193:256-263 (1975).

25. Dubois, A. B., S. Y. Botelho, G. N. Dedell, R. Marshall and J. H. Comroe, Jr. "A Rapid Plethysmographic Method for Measuring Thoracic Gas Volume: A Comparison with a Nitrogen Washout Method for Measuring Functional Residual Capacity in Normal Subjects," *J. Clin. Invest.* 35:322-326 (1956).

26. Boche, R. D., and J. J. Quilligan. "Effect of Synthetic Smog on Spontaneous Activity of Mice," *Science* 131:1733 (1960).

27. Gage, M. I., Y. Y. Yang, A. L. Cohen and J. F. Stara. "Alterations of Wheel Running Behavior of Mice by Automobile Fuel Emissions," paper presented at the American Psychological Association Meeting, September 1972.

28. Gage, M. I., Y. Y. Yang and J. F. Stara. "Automotive Exhaust Suppression of Mouse Activity Modified by a Manganese Additive, Intermittent Exposure and a Catalytic Converter," *Fed. Proc.* 33: 482 (1974).
29. Kehoe, R. A., W. F. Mackle, K. Kitzmiller and T. J. LeBlanc. "On the Effects of Prolonged Exposure to Sulfur Dioxide," *J. Ind. Hyg.* 14:159-173 (1932).
30. Terzioglu, M., and F. Emiroglu. "The Effects of Carbon Monoxide Inhalation on Respiratory Activity and Blood Acid Base Equlibrium of Normal and Chemoreceptorless Rabbits," *Arch. Physiol. Biochem.* 65:13-26 (1957).

CHAPTER 12

MUTAGENIC ACTIVITY OF
AIRBORNE PARTICULATE ORGANIC POLLUTANTS

James N. Pitts, Jr., Daniel Grosjean and Thomas M. Mischke

Statewide Air Pollution Research Center,
University of California
Riverside, California 92521

Vincent F. Simmon and Dennis Poole

Division of Life Sciences
Stanford Research Institute
Menlo Park, California 94025

Organic extracts of airborne particulate matter samples collected in 1975 and 1976 at 11 urban sites and one nonurban location in the Los Angeles area were tested for mutagenic activity using Ames' *Salmonella typhimurium* assay system.

Assay of 0.1-1 mg of the samples resulted in a 5- to 20-fold increase in *his*+ revertants above the background level, with the number of revertants per plate ranging from 0.07-0.51 per μg of organic carbon tested. Activity was observed with strains TA1537, TA1538 and TA98, which are reverted by frameshift mutagens but not seen with TA1535, which is reverted by base-pair substitution mutations. The activity of most samples (20 out of 23) was not enhanced when tested in the presence of metabolic activation (0.5 ml of liver S9 homogenate). Finally, assay of a size-resolved sample revealed that essentially all mutagenic activity was associated with organic species in particles of diameter 1.1 μ and less.

Since frameshift-type mutagenic activity without metabolic activation was found in all urban samples, these must contain mutagens other than

219

benzo(a)pyrene (BaP) and other carcinogenic polycyclic aromatic hydro-carbons (PAH) that require metabolic activation in the Ames test. We propose that certain of these directly active chemical mutagens may be formed in atmospheric reactions of ambient PAH with species present in photochemical smog, *e.g.,* ozone and nitrogen dioxide. Recently we have shown that such reactions can occur in simulated atmospheres with BaP deposited on a typical high-volume filter. Thus the question of the degree of reaction in the atmosphere versus that during collection on the filter becomes highly relevant to this and to previous studies of the carcinogenicity of ambient organic particulates.

INTRODUCTION

The tumorigenicity and/or carcinogenicity of extracts of the organic fraction of airborne particulate matter has been known for 35 years.[1,2] It was first reported for Los Angeles smog in 1954 by Kotin et al.[3] (with additional papers in 1956 and 1958[4,5]) and for seven U.S. cities (includ-ing Los Angeles) by Hueper and co-workers in 1962.[6] The subject has been reviewed by Epstein et al.[7] in "Particulate Polycyclic Organic Matter," published in 1972 by the National Academy of Sciences,[8] and by Hoffmann and Wynder[9] in 1977. These should be consulted for other references to the original literature.

Although the subject has been of interest for more than three decades, because of the great chemical complexity of organic particulates and because animal studies required to test various known and suspected carcinogens are time-consuming and expensive, experiments directed to establishing the chemical nature of the compounds responsible for the observed carcinogenicity have been relatively limited.[3-11]

Recently, microbiological assays for mutagenic activity have been developed by Ames and co-workers[12-15] based on the back mutation of modified *Salmonella typhimurium* strains in a histidine-limited medium. A good correlation (about 90%) between carcinogenic and mutagenic activity (and the absence thereof) has been reported by Ames et al.[14,15] for about 300 chemicals tested.

Therefore, with the clear understanding that mutagenic activity in bacterial assays does not necessarily imply occurrence of cancer in man, these relatively simple, fast and inexpensive assays may be used as a first approach for the screening of presumptive atmospheric carcinogens.[16,17] Indeed, in his address to the Seventh Annual Environmental Mutagen Society Meeting F. J. de Serres stated:

> "During the past year there have been two major developments that will have an enormous impact in the near future. The first was

the identification of many environmental and industrial chemicals as carcinogens and the widespread belief that a high percentage (80-90%) of human cancer is a result of such exposures. The second was a demonstration of a high correlation between carcinogenic and mutagenic activity in newly developed short-term tests for mutagenicity. The latter finding has enormous implications for the first, because it opens up the possibility to utilize the short-term tests not only to identify potential mutagens in our environment but potential carcinogens as well."[18]

As indicated above, the Ames test appears to be especially useful for testing complex mixtures such as particulates from polluted air samples in which several hundreds of chemicals have been identified.[9,19-21] and many still await identification. Thus, in 1975 we reported preliminary results showing for the first time mutagenic activity in the organic fraction of ambient atmospheric particulates using Ames' assay system.[16] We recently confirmed and extended our initial results to samples collected at nine locations in the Los Angeles area.[22] Subsequent to our 1975 study, mutagenic activity was reported in ambient particulate samples in Ohmuta and Fukuoka, Japan,[23] Buffalo, NY[24] and Berkeley, CA.[24]

Our study has been extended to include ambient samples collected at 12 locations in the California South Coast Air Basin. Results for these experiments and one exploring the mutagenic activity of the organic extracts of a size-resolved particulate air sample are presented here. A brief discussion of the possible modes of formation of certain of the chemical species responsible for such activity is also given.

EXPERIMENTAL PROCEDURES

Ambient particulate samples were collected in June-July 1975, December 1975 and September-October 1976, at a total of 12 locations in the California South Coast Air Basin (SCAB). These 12 sampling sites are shown in Figure 1.

In July 1975, one composite three-week sample was collected at each of the following four locations: Mountain View, a primary school located about 3 km south of Ontario Airport; Fontana, in the vicinity of an important power plant and a major steel mill; Santa Ana Canyon, about 40 km southwest (upwind) of Riverside on Highway 91; and Riverside, on the University of California campus. In December 1975, up to five consecutive samples were collected over periods of four to six days at the same four locations.

In September-October 1976, ambient particulate samples were collected over three-day periods at one nonurban site (Camp Paivika, elevation 5,320 ft in the San Bernardino Mountains) and eight urban locations:

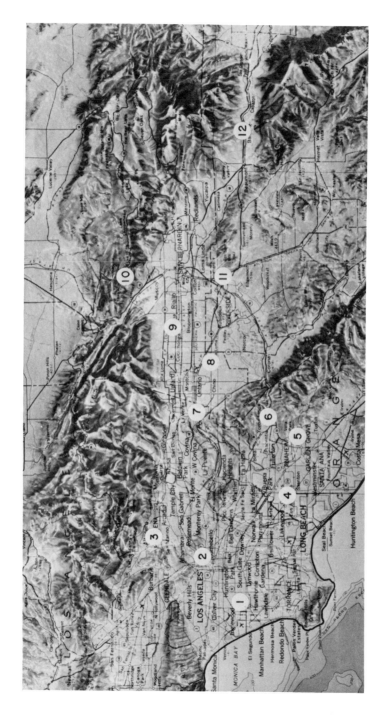

Figure 1. Map of the California South Coast Air Basin showing the 12 sampling sites: Lennox (No. 1), Los Angeles (2), Pasadena (3), Los Alamitos (4), Anaheim (5), Santa Ana Canyon (6), Pomona (7), Mountain View (8), Fontana (9), Camp Paivika (10), Riverside (11) and Banning (12).

Anaheim, Banning, Lennox, Los Angeles, Los Alamitos, Pasadena, Pomona and Riverside. Sampling locations and sampling periods are listed in Table I, as well as the total suspended particulates (TSP) and particulate organic carbon (POC) levels averaged over the corresponding sampling period. POC concentrations were determined using a Dohrman Model DC-50 Organic Carbon Analyzer.[25]

Samples were collected on Gelman type A-E glass fiber filters using high-volume samplers operating at constant flowrates of 40 or 60 cfm. The size-resolved sample was collected from April 4-8, 1977 in downtown Los Angeles using a five-stage Sierra Instruments Model 230 high-volume cascade impactor operating at 40 cfm. The nominal 50% cutoff sizes of the impactor were 8.2 μ, 3.6 μ, 2.1 μ, 1.1 μ and 0.65 μ.

The organic components were extracted by ultrasonication of the loaded filters with 200 ml of equal parts by volume mixture of methanol, benzene and dichloromethane (Burdick and Jackson, distilled-in-glass). The organic extracts were filtered and reduced to a measured volume of 1-3 ml using a Kuderna-Danish concentrator, then stored in the dark at 5°C in glass vials with Teflon®-lined caps prior to testing for mutagenic activity.

The concentrated organic extracts were tested for mutagenic activity by using the Ames *Salmonella typhimurium* assay system.[12-15] Each sample was tested twice at at least six doses (typically 0.1, 0.5, 1, 5, 10 and 50 μl) on five histidine-dependent strains of *Salmonella* (TA1535, TA1537, TA1538, TA98 and TA100) in the absence and presence of 0.5 ml of rat liver homogenate S9 metabolic activation system.[26,27] Appropriate positive and negative controls as well as blanks consisting of the solvent mixture and of extracts obtained from ultrasonication of the solvent mixture with unused filters and concentrated as indicated before were included in each assay.

To ensure that no artifact was created as a result of our concentration procedure, extracts were also concentrated by vacuum distillation of the solvent at ambient temperature. Mutagenic assay of extracts of the same sample concentrated by the two techniques (Kuderna-Danish and vacuum distillation) gave identical results, thus indicating that no positive (oxidation of an inactive chemical to an active one) or negative (degradation of a mutagenic compound into an inactive form) artifacts were created when concentrating the extract at high temperature (80°C) using the Kuderna-Danish concentrator. Solvents recondensed after use in the concentrator were also tested. The condensates showed no activity, thus indicating that evaporative losses of mutagenic chemicals initially present in the sample were insignificant.

---

*Registered trademark of E. I. du Pont de Nemours and Company, Inc., Wilmington, Delaware.

Table I. Collection of Airborne Particulate Matter Samples Tested for Mutagenic Activity. Summary of Sampling Locations, Sampling Periods, Total Mass Collected, Total Suspended Particulates (TSP) Concentrations, and Particulate Organic Carbon (POC) Concentrations

Sampling Site[a]	Sampling Period	Total Mass Collected (g)	TSP ($\mu g/m^3$)[b]	POC ($\mu g/m^3$)[b]	POC/TSP (%)
Summer 1975					
Mountain View	June 25-July 14	6.72	131	—	—
Riverside	June 25-July 14	3.63	114	—	—
Fontana	June 25-July 14	5.67	129	—	—
Santa Ana Canyon	June 25-July 14	4.17	113	—	—
Winter 1975					
Riverside	Dec. 2-5	0.70	157	—	—
	Dec. 10-15	1.08	132	—	—
	Dec. 15-19	0.34	66	—	—
	Dec. 19-24	0.82	95	—	—
Santa Ana Canyon	Dec. 14	0.54	112	—	—
	Dec. 4-10	1.33	142	—	—
	Dec. 10-15	0.53	63	—	—
Mountain View	Dec. 5-10	1.36	176	—	—
	Dec. 15-19	0.96	153	—	—
	Dec. 19-24	0.94	115	—	—
Fontana	Dec. 5-10	1.00	160	—	—
Summer 1976					
Camp Paivika	Oct. 4-6	0.14	27.8	2.9	10.4
Banning	Oct. 4-6	0.31	66	7.1	11.0
Pomona	Sept. 30 and Oct. 4-6	0.79	123	13.1	10.7
Riverside	Oct. 4-6	0.68	140	12.6	9.0
Pasadena	Sept. 29-Oct. 1	0.21	43.0	5.1	11.9
Los Angeles	Sept. 29-Oct. 1	0.19	39.8	5.6	13.9
Lennox	Sept. 25-27	0.14	28.0	2.6	9.3
Anaheim	Sept. 25-27	0.16	33.4	3.5	10.3
Los Alamitos	Sept. 22-24	0.40	82	7.1	8.7

[a] See Figure 1.
[b] Averaged over the corresponding sampling period.

RESULTS

Mutagenic assays conducted on the concentrated organic extracts of 24 airborne particulate samples collected at 12 sites in the SCAB resulted in the following findings:

1. All solvent and filter blanks were inactive (Table II). Although containing appreciable amounts of organic material (14.1 mg of organic carbon, POC = 2.9 μg m^{-3}), the sample collected at the nonurban site, Camp Paivika, was also inactive (Table III).

2. All 23 samples collected at urban sites exhibited mutagenic activity without requiring metabolic activation with strains TA1537, TA1538 and TA98, which are susceptible to frameshift mutations.

3. Typically, assay of 0.1-2 mg of airborne organic particulates resulted in a 5- to 20-fold increase in the number of histidine revertants per plate above the background of spontaneous revertants in strains TA1537, TA1538 and TA98, and up to a two-fold increase in strain TA100.

4. No activity was observed in any of the assays with strain TA1535, which is reverted by base-pair substitution mutations.

5. Dose-response curves were found to be approximately linear for at least the four highest doses of extract tested. Assay of some of the less active samples at the two lowest doses (0.05 and 0.1 μl) did not result in a significant increase in the number of revertants above the background number of spontaneous reversions. For a given dose of extract, strain TA98 was generally found to be the most sensitive, followed by TA1538 and TA1537 (Figure 2). Many, but not all, samples induced mutations in strain TA100, and the number of revertants per plate was always low compared to strains TA98, TA1538 and TA1537.

6. Addition of 0.5 ml of rat liver homogenate S9 did not increase the activity of the majority (20 out of 23) of active samples tested to date (Figure 3). In fact, addition of 0.5 ml of S9 in some cases resulted in a slight decrease in activity, as previously observed in this laboratory[16,22] and by Talcott and Wei[24] for an ambient air sample collected at Berkeley, CA. The dependence of the observed activity of ambient air samples and of pure compounds on S9 concentration is currently being investigated in our laboratory.

7. Three samples exhibited an increase in activity when tested in the presence of 0.5 ml of S9 liver homogenate: the activity of two of these samples, Mountain View (summer 1975) and Pomona, increased by 75% and 36%, respectively. The third sample (Fontana, summer 1975) exhibited a three-fold increase in activity (Figure 4).

8. From the linear region of the dose-response curves and the measured amounts of organic carbon (POC) in the extract tested, we have calculated the mutagenic activity (number of revertants per plate) per microgram of particulate organic carbon tested. As shown in Table IV, the activity of the samples tested to date with strains

Table II. Control Experiments and Assay of Solvent and Filter Blank

Compound	Metabolic Activation[a]	Microliters of Sample Added per Plate	Histidine Revertants per Plate				
			TA1535	TA1537	TA1538	TA98	TA100
Negative Control (Dimethylsulfoxide)	−	50	11	2	12	10	59
	+	50	10	3	15	14	71
Positive Controls							
β-Propiolactone	−	50 μg	117				
9-Aminoacridine	−	100 μg		391			
2-Anthramine	−	10 μg	17	3	6	15	63
	+	10 μg	449	208	1953	1745	1895
Solvent and filter blank[b]	−	0.05	13	1	10	13	47
	−	0.10	15	10	4	10	51
	−	0.50	7	1	15	14	54
	−	1.00	19	3	13	9	57
	−	5.00	16	3	9	7	54
	−	10.00	18	3	12	9	62
	+	0.05	13	4	12	16	59
	+	0.10	15	2	8	9	60
	+	0.50	15	6	9	13	72
	+	1.00	12	2	13	12	64
	+	5.00	3	3	9	16	69
	+	10.00	9	6	13	15	53

[a] With (+) or without (−) 0.5 ml of rat liver S9 added.
[b] Blank prepared by extraction of a blank Gelman A-E glass filter with the benzene-methanol-dichloromethane solvent mixture (see text).

Table III. Assay of the Organic Extract of a Particulate Sample Collected at a Nonurban Site (Camp Paivika) (No Mutagenic Activity was Detected in this Sample)

Compound	Metabolic Activation[a]	Microliters of Sample Added per Plate	Histidine Revertants per Plate				
			TA1535	TA1537	TA1538	TA98	TA100
Negative Control (Dimethylsulfoxide)	−	50	10	2	11	17	63
	+	50	8	4	19	16	74
Positive Controls							
β-Propiolactone	−	50 µg	239	–	–	–	189
2-Anthramine	−	10 µg	–	–	13	18	–
	+	10 µg	–	–	1676	2179	–
Camp Paivika[b]	−	0.05	7	12	6	14	51
	−	0.10	13	16	5	19	42
	−	0.50	15	9	10	23	49
	−	1.00	9	7	6	10	61
	−	5.00	13	10	10	28	44
	−	10.00	7	15	5	29	36
	+	0.05	10	4	11	14	69
	+	0.10	8	3	10	24	64
	+	0.50	8	2	12	22	85
	+	1.00	13	2	7	32	66
	+	5.00	12	9	8	19	83
	+	10.00	9	11	16	24	69

aWith (+) or without (−) 0.5 ml of rat liver S9.
bSee Table I for sampling conditions, mass collected, and TSP and POC data.

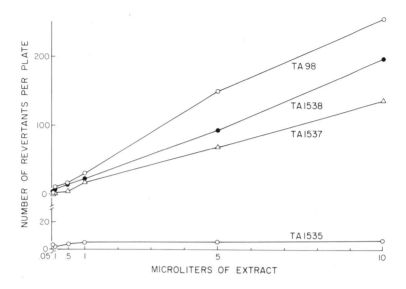

Figure 2. Dose-response curves for a sample collected in Los Angeles, California, September 29-October 1, 1976 (see Table I) and tested without metabolic activation.

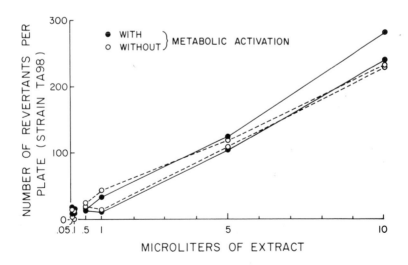

Figure 3. Dose-response curves for a sample collected in Riverside, California, October 4-6, 1976. The sample was tested twice without (open circles) and with (full circles) metabolic activation (0.5 ml of S9).

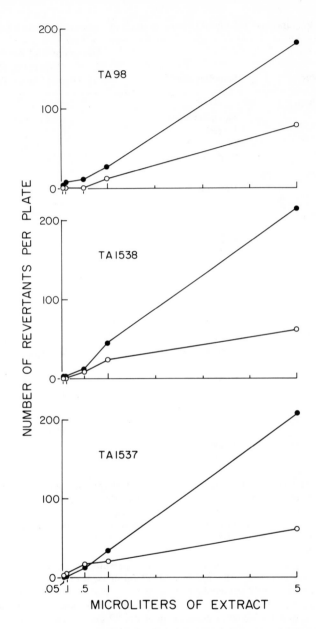

Figure 4. Dose-response curves for a sample collected in Fontana, California, June 25-July 14, 1975 and tested without (open circles) and with (full circles) metabolic activation (0.5 ml of S9 with frameshift strains TA98 (top), TA1538 (center) and TA1537 (bottom).

Table IV. Mutagenic Activity of Organic Extracts of Airborne Particulate Samples[a]

Sampling Period[b]	Sampling Site	Number of His[+] Revertants per Plate per μg of Organic Carbon Tested
Summer 1975		
	Mountain View	0.08 (0.14)[c]
	Riverside	0.17
	Fontana	0.14 (0.41)
	Santa Ana Canyon	0.18
Winter 1975		
	Riverside	0.12
		0.12
		0.29
		0.17
	Santa Ana Canyon	0.21
		0.07
		0.14
	Mountain View	0.16
		0.24
		0.26
	Fontana	0.51
Summer 1976		
	Camp Paivika	No activity
	Banning	0.15
	Pomona	0.25 (0.34)
	Riverside	0.19
	Pasadena	0.30
	Los Angeles	0.44
	Lennox	0.24
	Anaheim	0.21
	Los Alamitos	0.10

[a] Data are for strains TA1538 (Summer 1975 and 1976) and TA1537 (Winter 1975), *without metabolic activation*, after subtraction of the spontaneous and solvent backgrounds. Particulate organic carbon content was measured for all 1976 samples and was estimated for the 1975 samples to be 10% of the total mass collected based on the 1976 data (average 10.57%, see Table I).

[b] See Table I.

[c] Data with metabolic activation are indicated in parentheses for the three samples whose activity increased in the presence of 0.5 ml of rat liver homogenate S9. Only one concentration of S9 was employed in these initial studies (10% v/v or 0.1 ml liver homogenate in a total volume of 1 ml of the S9 mix). As the observed mutagenicity of any given activatable mutagen depends upon the S9 concentration used, S9 dose-response curves should be determined (whenever sample size permits) to avoid misleading results due to over-metabolism, or "S9 suppression."

TA1537 or TA1538 in the absence of metabolic activation varied from 0.07 to 0.51 revertants per plate per μg of organic carbon.

9. Assay of organic extracts of a size-resolved sample collected in downtown Los Angeles revealed that all mutagenic activity was associated with organic species present in particles of diameter 1.1 μ or less (Table V). This is consistent with the well-documented distribution of organic particulate pollutants such as benzo[a]pyrene (BaP) with respect to particle size.[19,28,29]

Table V. Mutagenic Activity of a Size-Resolved Particulate Sample Collected in Los Angeles, California, April 4-8, 1977

Size Range (μ)	Mass Collected (mg)	Mutagenic Activity (Strain TA 98)
>8.2	91.4	No activity (NA)
3.6 - 8.2	100.6	NA
2.1 - 3.6	67.3	NA
1.1 - 2.1	82.5	NA
0.65 - 1.1	128.3	2 x[a]
<0.65	419.0	20 x(−), 30 x(+)[b]

[a] nx = n times increase in activity above the spontaneous and solvent backgrounds.
[b] With (+) and without (−) metabolic activation (0.5 ml of rat liver S9).

DISCUSSION AND CONCLUSIONS

Although mutagenic activity of particulate organic matter collected from urban air and determined by the Ames test had not been reported prior to our initial studies,[16] as we pointed out earlier, the carcinogenic activity of organic particulate matter collected in the Los Angeles atmosphere has been well established,[3,7,10,11] as it has for other urban areas.[6,9] Therefore, our findings of mutagenicity in all urban samples tested is not surprising. Indeed, in a sense they not only confirm the utility of the Ames test in studying the microbiological activity of organic particulate matter but also suggest that such a test can provide valuable clues to the chemical nature of the organic compounds responsible for the carcinogenic activity in test animals.

Two major classes of primary pollutants found in airborne organic particulates have been proposed to be responsible for the observed induction of cancer in animals—the polycyclic aromatic hydrocarbons (PAH) and their nitrogen-containing analogs.[8,9,30] For example, Epstein et al.[7] attributed the carcinogenicity of the neutral fraction of organic particulates to the presence of such PAH as benz[a]anthracene and benzo[a]pyrene, as well as the presence of a polynuclear carbonyl compound, 7H-benz[d,e]anthracen-7-one. A number of other primary pollutants carcinogenic to animals have been identified in ambient air, including numerous PAH and heterocyclic compounds such as benzocarbazoles and benz- and dibenzacridines.[9,20,21,31]

However, since most PAH (including benzo[a]pyrene) require metabolic activation to induce frameshift mutations in the Ames assay system,[14,15] our results suggest that BaP and other PAH cannot account for more than a relatively small fraction of the mutagenicity of our samples. They may, however, contribute to the increase in activity of three of the samples when tested in the presence of S9. Our results are consistent with those of Gordon and co-workers,[11] who reported that airborne particles from Los Angeles contained potentially carcinogenic material not extracted by the conventional benzene treatment but soluble in methanol. The benzene extract had 100-1,000 times the cell transformation activity attributable to its measured BaP content alone. Furthermore, the activity of the methanol extract was comparable to that of the benzene extract, despite the fact that the former contained 30 times less BaP than the latter.

The negative results obtained in all cases with strain TA1535 suggest that alkylating agents which induce base-pair substitution mutations are probably not present in our samples. Furthermore, based on vapor pressure considerations, most alkylating agents known or suspected to be present in polluted air (for example vinyl chloride, epoxides, N-nitrosamines) are expected to be in the gas phase rather than the particulate phase and, therefore, would not be collected using our sampling procedure.

While a number of organic compounds may contribute to the mutagenic activity we observe in particulate matter (including carboxylic acids, nitrate esters, polyfunctional oxygenates, etc.[9,19]) we feel that important candidates may be the products of the oxidation or nitration of ambient PAH (including BaP) by constituents of photochemical smog. Thus, oxygenated fractions of ambient particulates have been found to be carcinogenic.[9-11]

Our observation that the mutagenic activity is found almost exclusively in the smallest size range (< 0.65 μm) is consistent with the fact that in urban aerosols BaP is associated with fine particles of a diameter ~0.15 μm and less.[28,29] Such particles, of course, are readily entrained in the pulmonary region.[29] In this small size range, oxidation of surface BaP and other PAH should occur readily in the highly oxidizing atmosphere characteristic of photochemical smog.

Oxidative processes may include the reactions of benzo[a]pyrene and other PAH with ozone, NO_2 and peroxyacetylnitrate (PAN) to produce directly active mutagens (active without S9) from compounds otherwise requiring metabolic activation. Indeed, we have found[32-35] that when BaP was exposed to ambient photochemical smog on a filter, the resulting oxidation products were directly active when tested with the frame-shift strains of *Salmonella*. Furthermore, exposure of BaP on a filter to 0.25 ppm of NO_2 in air (with a trace of HNO_3) resulted in the formation of three directly active mononitrobenzo [a] pyrene isomers.[32-35] *

Whether these reactions, which readily occur on the filter surface, also take place in the atmosphere on the surface of airborne particles, is a matter of conjecture. Conversely, if the reactions listed above are due to a large extent to filter "artifacts," results of previous studies reporting carcinogenic activity in samples of airborne particulate matter collected on filters might be questionable. Work is now in progress in our laboratory to investigate these issues and to further elucidate the nature of those organic pollutants responsible for the mutagenic activity of airborne particulate matter.[32-36]

ACKNOWLEDGMENTS

Useful discussions with Professor W. L. Belser, Jr., Dr. G. B. Knudson and Mr. P. M. Hynds, Department of Biology, University of California, Riverside, and Drs. K. Van Cauwenberghe and J. Schmid of the Statewide Air Pollution Research Center are gratefully acknowledged. This work was supported by National Science Foundation-Research Applied to National Needs Grants No. AEN 73-02904 A02 and ENV 73-02904 A03, and by Stanford Research Institute internal research and development funds. The contents of this chapter do not necessarily reflect the views and/or policies of the NSF-RANN, nor does mention of trade names or commercial products constitute endorsement or recommendation for use.

---

*This level is the California air quality standard for NO_2; it is often exceeded. Recently, we have shown that exposure of BaP on a filter to 0.10 ppm O_3 in air also results in the facile formation of directly mutagenic species. This concentration is the California air quality standard for this pollutant. The BaP-O_3 reaction may well be a major oxidative pathway for the destruction of ambient PAH and the generation of at least part of the directly mutagenic species found in ambient organic particulates.

REFERENCES

1. Leiter, J., M. B. Shimkin and M. J. Shear. *J. Nat. Cancer Inst.* 3: 155 (1942).
2. Leiter, J., and M. J. Shear. *J. Nat. Cancer Inst.* 3: 455 (1943)..
3. Kotin, P., H. L. Falk, P. Mader and M. Thomas. *Arch. Ind. Hyg.* 9:153 (1954).
4. Kotin, P., H. L. Falk and M. Thomas. *Cancer* 9:905 (1956).
5. Kotin, P., H. L. Falk and C. J. McCammon. *Cancer* 11:473 (1958).
6. Hueper, W. C., P. Kotin, E. C. Tabor, W. W. Payne, H. L. Falk and E. Sawicki. *Arch. Pathol.* 74:89 (1962).
7. Epstein, S. S.,,N. Mantel and T. W. Stanley. *Environ. Sci. Technol.* 2:132 (1968).
8. Particulate Polycyclic Organic Matter, National Academy of Sciences, Washington, DC, 1972.
9. Hoffmann, and E. L. Wynder. "Organic Particulate Pollutants–Chemical Analysis and Bioassays for Carcinogenicity" in *Air Pollution*, Vol. II, 3rd ed., A. C. Stern, Ed. (New York: Academic Press, 1977).
10. Wynder, E. L., and D. Hoffmann. *J. Air Poll. Control Assoc.* 15: 155 (1965).
11. Gordon, R. J., R. J. Bryan, J. S. Rhim, C. Demoise, R. G. Wolford, A. E. Freeman and R. J. Huebner. *Int. J. Cancer* 12:223 (1973).
12. Ames, B. N., W. E. Durston, E. Yamasaki and F. D. Lee. *Proc. Nat. Acad. Sci.,* U.S. 70:2281 (1973).
13. Ames, B. N., F. D. Lee and W. E. Durston. *Proc. Nat. Acad. Sci.,* U.S. 70:782 (1973).
14. McCann, J., E. Choi, E. Yamasaki and B. N. Ames. *Proc. Nat. Acad. Sci.,* U.S. 72:5135 (1975).
15. McCann, J., and B. N. Ames. *Proc. Nat. Acad. Sci.,* U.S. 73:950 (1976).
16. Pitts, J. N., Jr. "Chemical Transformations in Photochemical Smog and Their Applications to Air Pollution Control Strategies," Second Annual Progress Report, National Science Foundation-Research Applied to National Needs Grant No. AEN73-02904 A02, September (1975), p. v-8.
17. Commoner, B. U.S. Environmental Protection Agency Report No. EPA-600/1-76-022, Washington, DC (1976).
18. de Serres, F. J. *Mut. Res.* 38:355 (1976).
19. Grosjean, D. "Aerosols," *Ozone and Other Photochemical Oxidants,* (Washington, DC: National Academy of Sciences, 1977), p. 45.
20. Cautreels, W., and K. Van Cauwenberghe. *Atmos. Environ.* 10:447 (1976).
21. Cautreels, W., and K. Van Cauwenberghe. *Sci. Total Environ.* 8:79 (1977).
22. Pitts, J. N., Jr., D. Grosjean, T. M. Mischke, V. F. Simmon and D. Poole. *Toxicol. Lett.* 1:65 (1977).
23. Tokiwa, H., H. Takeyoshi, K. Morita, K. Takahashi, N. Soruta and Y. Ohnishi. *Mut. Res.* 38:351 (1976).
24. Talcott, R., and E. Wei. *J. Nat. Cancer Inst.* 58:449 (1977).
25. Grosjean, D. *Anal. Chem.* 47:797 (1975).

26. Kier, L. D., E. Yamasaki and B. N. Ames. *Proc. Nat. Acad. Sci.*, U.S. 71:4159 (1974).

27. Ames, B. N., J. McCann and E. Yamasaki. *Mut. Res.* 31:347 (1975).

28. Kertesz-Saringer, M., E. Meszaros and T. Varnkonyi. *Atmos. Environ.* 5:429 (1971).

29. Natusch, D. F. S., and J. R. Wallace. *Science* 186:695 (1974).

30. Sawicki, E. *Arch. Environ. Health* 14:46 (1967).

31. Sawicki, E., S. P. Pherson, T. W. Stanley, J. Meeker and W. C. Elbert. *Int. J. Air Water Poll.* 9:515 (1965).

32. Pitts, J. N., Jr. "Photochemical and Biological Implications of the Atmospheric Reactions of Amines and Benzo(a)pyrene," Discussions of the Royal Society on "Pathways of Pollutants in the Atmosphere," London, November 3-4, 1977; *Philosophical Transactions of the Royal Society*, In press.

33. Pitts, J. N., Jr., K. Van Cauwenberghe, D. Grosjean, J. Schmid, D. Fitz, W. Belser, Jr., G. Knudson and P. Hynds. "Reactions of Polycyclic Aromatic Hydrocarbons in Real and Simulated Atmospheres: Facile Formation of Mutagenic Nitroderivatives of Benzo(a)pyrene and Perylene," *Science* 202:515 (1978).

34. Pitts, J. N., Jr., W. L. Belser, Jr., K. Van Cauwenberghe, D. Grosjean, J. Schmid, D. R. Fitz, G. B. Knudson, and P. Hynds. "Chemical and Microbiological Studies of Mutagenic Pollutants in Real and Simulated Atmospheres," presented at the EPA-sponsored Symposium "Application of Short-Term Bioassays in the Fractionation and Analysis of Complex Environmental Mixtures," Williamsburg, VA, February 21-23, 1978.

35. Pitts, J. N., Jr., D. Grosjean, K. Van Cauwenberghe and W. L. Belser, Jr. "Mutagenic Activity of Nitrogenous Pollutants in Ambient and Simulated Atmospheres," Division of Environmental Chemistry, Symposium on Chemical and Biological Implications of Nitrogenous Air Pollutants, 175th National ACS Meeting, Anaheim, CA, March 12-17, 1978.

CHAPTER 13

BIOPHYSICAL STUDIES OF THE EFFECTS OF ENVIRONMENTAL AGENTS ON THE PLASMA MEMBRANE OF INTACT CELLS

J. R. Rowlands, C. Allen-Rowlands,
A. Noyola and J. Padilla

Southwest Foundation for Research and Education
San Antonio, Texas 78284

INTRODUCTION

Environmental agents, after entering into the host, are capable of distribution throughout the system. While the physicochemical nature of the interaction of a given agent with different cell types will be the same, *i.e.,* covalent binding or intercalation in membrane lipids, etc. the affinity of the agent might differ markedly between cell types due to their wide differences in membrane composition. Thus, it is extremely difficult to predict *a priori* which biological effect might be most significant for a given pollutant. Due to the great variability observed both among animal species and within a given species—even when inbred species are studied—it becomes exceedingly time-consuming and difficult to assess the *in vivo* pharmacological effect of environmental agents at concentrations that approximate those in the environment. However, an understanding of the biochemical and biophysical mechanisms by which environmental agents interact with different cell types *in vitro*, when correlated with the corresponding physiological response of these cell types, would greatly enhance our ability to predict the potential pharmacology of environmental agents and help in better design of whole-animal experiments.

That gases such as sulfur dioxide (SO_2) and nitrogen dioxide (NO_2) are rapidly distributed throughout the system has been demonstrated by radioactive tracer studies. Goldstein *et al.*[1] have recently reported that rhesus monkeys, after inhalation of air mixtures containing 0.30-0.91 ppm NO_2 labeled with tracer amounts of $^{13}NO_2$, retained 50-60% of the

237

inspired NO_2 in their lungs during quiet respiration. The NO_2 was retained in the lungs for prolonged intervals after exposure but was then transported out of the lungs either as NO_2 or a derivative of NO_2 via the blood stream. Several years ago Rowlands and Gause[2] found by electron spin resonance techniques that a well-defined paramagnetic complex could be identified in the blood of rats that had been exposed *in vivo* to 1-3 ppm NO_2. We have recently conducted experiments in which we examined the temporal and regional distribution of $^{35}SO_2$ following a 30-minute exposure of four male Sprague-Dawley rats to 5 mci $^{35}SO_2$. We found that while the radioactivity in the lungs remained relatively constant throughout the experiment, initially there was a large uptake of $^{35}SO_2$ by the adrenal glands compared to other tissues examined. The pituitary gland and testes took up a significant amount of radioactivity with time. The fact that $^{35}SO_2$ crosses the blood brain barrier was demonstrated by the fact that radioactivity was detected both in the pituitary gland and in other brain parts examined.

Environmental agents that can interact with membrane components of the neuroendocrine system (*i.e.*, adrenal, pituitary, CNS) have the potential to alter neurohormonal mechanisms controlling and/or modulating the dynamics of the endocrine systems. It has been demonstrated that many disease states in man and animals are associated with decreased or increased hormonal sensitivity. Since hormone-receptor interaction is the first step in the action of the hormone, the disruption of either CNS or target organ cellular membranes by environmental agents could significantly disrupt the normal hormonal cascade of the endocrine axis. The observation that SO_2 interacts at the cellular level with endocrine and CNS organs could be particularly important for understanding the effects of chronic exposure of the population to this pollutant.

Despite extensive research conducted on the health effects of both of these environmental contaminants, the fundamental question of their precise mechanism of action remains ill defined. For example, in *in vitro* studies of the mechanism of action of NO_2 with phospholipid systems, Rowlands and Gause[2] showed by electron spin resonance techniques that several free radical species were formed. The chemical nature of these free radicals was established and reported. However, in similar *in vitro* experiments utilizing intact alveolar macrophages (AM), these authors[3] were unable to detect the presence of free radical products. In the case of SO_2 this pollutant has generally been considered to immediately convert to bisulfite and all subsequent reactions have been attributed to the chemistry of this anion. Ths possibility that SO_2 might react via free radical reactions was not even considered until Eickenroht *et al.*[4,5] showed that under *in vitro* conditions, which simulated the *in vivo* situation, SO_2 forms a molecular complex with electron donor molecules such as tryptophan and can be induced to accept an electron to form the radical anion $SO_2{}^-$ from light of wavelengths entering the atmosphere. Furthermore, this radical anion is produced by naturally occurring biological reducing agents under both acid and alkaline conditions.[6]

In aqueous solution SO_2 and NO_2 are known to react according to the following equations:

$$SO_2 + xH_2O = SO_2\ xH_2O$$
$$SO_2\ xH_2O = H_2SO_3 \quad K \ll 1$$
$$SO_2 \cdot xH_2O = HSO_3{}^{-1}\ aq + (x - 2)\ H_2O$$
$$\frac{(HSO_3{}^-)\ (H^+)}{(SO_2) - (HSO_3{}^-) - (SO_3{}^=)} = 1.3 \times 10^{-2}$$

Whereas free sulfurous acid is unstable, salts of the acid, namely the metal bisulfites, do exist. Since the net result of this reaction is the release of protons, unless the buffering capacity of the medium is adequate, an acidic environment will result. Thus, due to the above reaction scheme, although the reactions of SO_2 typically have been assumed to be those of the bisulfite anion, there exists the distinct possibility that the SO_2 gas hydrate may also exist in the biological environment. Since the particularly effective electron accepting properties of the SO_2 molecule have recently been reported,[4] even small concentrations of SO_2 in equilibrium with the bisulfite anion could have a significant effect on the overall biological effects of this gas.

The aqueous solution chemistry of NO_2 is generally considered to be due to its reactive products with water, namely:

$$2\ NO_2 + H_2O = HNO_3 + HNO_2$$
$$3\ HNO_2 = HNO_3 + 2\ NO + H_2O$$

Nitrous acid (HNO_2) is known to possess both oxidizing and reducing capabilities. Thus, just as in the case of SO_2, the net result of NO_2 inhalation into the lung is the creation of an acid environment, which may or may not be compensated for by its natural buffering capacity.

Much of the *in vivo* research that has been conducted on the health effects of these two very important pollutants has been concerned with their effects on lung defense systems, with particular reference to their effect on functional properties of the alveolar macrophage. The lungs are constantly exposed to the external environment with its variable content of irritants and infectious agents. Agents deposited in or below the region of the respiratory bronchioles are phagocytized by the AM. The specific capacity of the AM to perform its task is subject to many factors. Gaseous air pollutants have been shown to affect the functional state of the cells.[7] Phagocytosis is an energy-dependent process, and plasma membrane ATPase has been suggested to act as a mechanoenzyme making phagocytosis possible through the conversion of chemical energy in the form of ATPase to the mechanical energy required for attachment and ingestion.[8] Since the bulk of the cellular ATPase activity is located in the plasma membrane of the AM,[9] the enzyme is easily accessible to inhaled pollutants. Furthermore, the activity of ATPase and other membrane-bound enzymes is well known to depend on the fluidity of membrane lipids. This chapter will focus on the use of the electron spin resonance spin label technique to study the effects of exogenous agents on the plasma membrane of intact cells.

Since the plasma membrane of the cell separates the cell from its environment, it is logical to suppose that it is the first site of attack of pathological agents whether physical, chemical or biological. The integrity of the exterior surface of the plasma membrane is essential for many biological phenomena. For example, phagocytosis, pinocytosis, cell recognition, antigen binding, etc., are essentially surface phenomena. Similarly, the specificity of endocrine-related cells that respond to peptide hormones relies on the integrity of the hormone receptor binding proteins, which are, in general, integral glycoproteins lying in the exterior surface of the plasma membrane. At present, the mechanism has not been elucidated whereby hormone-receptor interaction induces plasma membrane structural changes, which facilitate the activities of the adenyl cyclase in the target endocrine cell.

Much of our current knowledge of the dynamic properties of biological membranes has been obtained through the use of the electron spin resonance spin label technique. Although the pioneering work in this area has been conducted on model membranes or membrane fragments, we will describe biophysical studies conducted to examine the effect(s) of environmental contaminants on intact mammalian cells. A recent study by the spin label technique examined the effect of hormone receptor interactions on changes in intrinsic plasma membrane protein conformation, bulk fluidity of membrane lipids and phospholipid halo distribution.[10] By choosing the appropriate spin label, it was possible to monitor conformational changes of cell surface proteins and to measure changes in phospholipid bulk fluidity at different depths into the outer membrane monolayer. By appropriately designed experiments one could measure both intrinsic membrane protein conformational changes at different depths into the membrane and phospholipid distributional changes as a result of the interaction of the peptide hormone, adrenocorticotrophin (ACTH) with Y-1 adrenal cell membranes. Subsequent to these studies, further biophysical studies have been conducted to examine the effects of SO_2 on this cell system; the results will be discussed in this presentation. The results from similar biophysical studies designed to elucidate the effects of SO_2, NO_2 and pH on the plasma membrane of the alveolar macrophage of Fisher rats will also be described.

There has been considerable interest in recent years in determining the nature of the differences in cell surface properties of normal and chemically or virally transformed cells. For example, several investigators have shown that transformed Balb 3T3 cells agglutinate more rapidly than the normal counterpart.[11-14] This ability of lectins to induce clustered arrangements of receptor sites on transformed mouse fibroblasts has been demonstrated repeatedly.[11,15,16] The fluidity of membrane lipids has often been suggested as a possible factor that contributes to the differences in mobility of lectin-binding glycoproteins on the surfaces of normal and transformed cells. Differences in fatty acid composition[16] are small for both normal and transformed mouse fibroblasts. However, the flexibility of lipid chains is not always determined by chemical composition alone, but may also be affected by the presence of proteins in the intact membrane. Two opposing reports have been published on the question of fluidity properties of normal and transformed mouse fibroblasts. Barnett et al.,[17] by the spin label studies, report significant differences in order parameter (S) between the normal and transformed cells. They observe a significantly decreased order parameter for the transformed cells indicative of greater lipid fluidity. Gaffney,[18] in a similar study, reported that she could observe no detectable differences in order parameter. Barnett et al.[17] conducted their measurements at 33°C whereas Gaffney[18] made her measurements at 37°C. Preliminary ESR studies conducted in this laboratory will be presented in which we have detected differences in the plasma membrane of the contact-inhibited Balb 3T3 mouse fibroblast cells versus the chemically transformed Balb 3T3 cell system.

EXPERIMENTAL PROCEDURES

Cell Maintenance and Preparation

Y-1 Adrenal Tumor Cells

Stock monolayer cultures of mouse adrenal cortex (Y-1) clone obtained from American Type Culture Collection were maintained in T-75 plastic tissue culture flasks (Falcon) in Ham's nutrient mixture, F-10 medium supplemented with 15% horse serum and 2.5% fetal calf serum at 37°C. For experimentation, T-30 culture flasks were seeded with 4×10^5 cells and allowed to grow to near the stationary phase (10-14 days after seeding). Confluent cultures were washed twice with 3.0 ml of incubation buffer (125 mM Na$^+$; 4.8 mM K$^+$; 120.8 mM CL$^-$; 20 mM hydroxyethane piperazine sulfuric acid (HEPES); 2.4 mM Ca^{++}— *i.e.*, HEPES buffer) at pH 6.9. The cells were then incubated in the presence or absence of ACTH (60 mU/ml) and varying concentrations of NaHSO$_3$ for selected intervals at 37°C on a rotary shaker. At the end of the incubation, the medium was removed and saved for steroid and/or cAMP analysis if required. The attached cells were then washed twice with HEPES buffer and scraped into a 2.5-ml HEPES with a rubber policeman. The cells were transferred to a 15-ml plastic centrifuge tube and the incubation flask washed 4 times with 2.5 ml HEPES. The cells were centrifuged for 2 min at 900 x g and the supernatant decanted. The cells were then resuspended in 2.6 x 10^{-5} M concentration of desired spin label and incubated 1-10 min at 37°C. The cells were centrifuged at room temperature (RT) for 2 min at 900 x g, the supernatant discarded and the cells resuspended in 15 ml HEPES (RT) and recentrifuged. All the supernatant was carefully removed and the cells transferred to a Pasteur pipette and the tip sealed. The spin-labeled cells were positioned in the pipette and in such a manner that they would be within the active region of the ESR cavity. The spin labeling procedure was done on individual samples and was completed within 12-25 min of termination of the incubation procedure, depending on the time of incubation with the spin label. ESR measurements were made with a Varian E-4 EPR spectrophotometer (Varian Associates, Palo Alto, CA). Sample temperature was controlled by a Varian variable temperature attachment. Spin labeling procedures for the Balb 3T3 and macrophage experiments were performed in an essentially identical manner.

Balb 3T3 Cells

The cells used for the study were received from Dr. Takeo Kakunaga, who is presently working at the National Cancer Institute (NCI) in Bethesda, MD. The Balb 3T3 I-13 cells were nontreated cells while the Balb NQT cells had been chemically transformed by 4 nitroquinoline oxide. The two types of cells showed distinct morphological differences. The Balb 3T3 I-13 clone was contact inhibited and clearly grew in an epitheloid-like monolayer; however, a few areas of overgrowth were exhibited. The Balb NQT grew in a flowing, fibrous pattern and on the edges of colonies the cells clearly showed a criss-crossing pattern. The cells were passaged into sterile Corning T-25 and twice weekly were fed with MEM Earle's salt media containing 10% fetal calf serum. The cells were passaged at a specified dilution into Corning sterile T-75 flasks so that when the cells were analyzed, there were approximately 8 million cells per flask. The cells were removed from the flask wall with a rubber policeman and incubated with 2.6×10^{-5} M concentration of 5-doxyl stearic acid spin label in MEM without fetal calf serum and transferred to a Pasteur pipette for ESR analysis described for the Y-1 adrenal tumor cell.

Isolation of Alveolar Macrophages

Fisher rats were sacrificed by peritoneal injection of an anesthetic (Diabutol) overdose, followed by exsanguination at the bifurcation of the abdominal aorta. Subsequent to sacrifice, the lungs were excised from the thoratic cavity and all extraneous tissue, *i.e.*, heart and fasia, was removed. To harvest macrophages, the lungs were lavaged 4-6 times with a volume of $37°C$ physiological (0.85%) saline approximately equal to the predetermined lung volume to avoid distention and subsequent fracture of capillaries. Macrophages from 2-3 animals were pooled, counted and aliquoted such that there was 10^6 cells per 40-ml centrifuge tube. The volume was adjusted to 30 ml with phosphate-buffered saline (PBS) and an aliquote of stock 5-doxyl stearic acid spin label added to reach a final concentration of 2.6×10^{-5} M spin label. The cells were incubated for 15 min at $4°C$, centrifuged at 90 x g and resuspended in PBS with or without added contaminant. Following centrifugation at 900 rpm for 4 min, the supernatant was removed and the cells transferred to a Pasteur pipette for ESR analysis.

Materials

Spin-labeled compounds were purchased from Syva Co., Palo Alto, CA. They were nitroxides (N-oxyloxazoledine paramagnetic radicals) attached to stearic acid $I_{(12,3)}$, $I_{(1,14)}$, or containing an isothiocynate group, which is capable of covalently binding to primary amine or hydroxyl groups of proteins.

Chemical Name	Abbreviations
4-Isothiocyanato-2,2,8,6-tetramethylpiperidinooxyl	I_{103} $\equiv$ spin label I
2-(3-Carboxylpropyl)-4,4-dimethyl-2-tridecyl-3-oxazolidinyloxyl	$I_{12,3}$ $\equiv$ 5-doxyl stearic acid
2-(14-Carboxytetradecyl)-2-ethyl-4,4-dimethyl-3-oxazolidinyloxyl	$I_{1,14}$ $\equiv$ 16-doxyl stearic acid

ACTH (Cortrosyn, 1-24 corticotrophin) was obtained through the generosity of Organon Inc. Y-1 adrenal tumor cell extracellular cAMP was assayed by the radioimmunoassay technique of Harper and Brooker.[19] The radio-labeled antigens and antibodies were purchased from Schwartz Mann. Extracellular steroid levels were assayed fluorimetrically using a modification of the technique described by Vernikos-Danellis et al.[20] ATPase activity of the alveolar macrophage membranes was assayed by the method of Godin and Schrier,[21] and protein determinations were made by the method of Lowry.[22]

Electron Spin Resonance Spin Label Techniques

Three classes of useful parameters are present in nitroxide spin label spectra—the g tensor, the nitrogen hyperfine coupling tensor and the widths of the individual ESR lines. From their combinations, information can be obtained about the orientation and mobility of the nitroxide.

Spin labels such as 4-Isothiocyanato-2,2-8,6-tetramethylpiperidinooxyl have been designed to covalently bind to specific primary amine or hydroxyl groups on membrane proteins. In these cases, changes in the observed ESR spectra are indicative of protein conformational changes.

In ESR studies of the plasma membrane lipids, a "reporter" group, usually a nitroxide containing an unpaired electron, is attached to one of the carbons of a stearic acid molecule. Since the spin-labeled stearic acids intercalate into the lipid bilayer to lie parallel to the phospholipid side chains, their spectroscopic behavior is characteristic of the order and structure displayed by the lipids constituting the plasma membrane in question.

The ESR spectra of fatty acid spin labels in lipid bilayers are determined by the rapid anisotropic motion of the fatty acid molecules,[23,24] which are anchored with their polar heads in the membrane-water interface. The alkyl chains can rotate about their long axis, which, in turn, precess about the normal to the membrane surface. Superimposed on these rigid body motions are intrachain motions involving transgauche isomerizations about the C-C bonds. Four ESR parameters were investigated in these studies.

Order Parameter and Correlation Time Measurements

The "fluidity" of a membrane can be characterized by the so-called order parameter $S^{23,24}$ defined by

$$S = (T_{\parallel}' - T_{\perp}')/T_{zz} - T_{xx})$$

where $T_{zz} - T_{xx} = 25$ gauss.[25] S determines the average orientation of
the long axis of the $2p\pi$ orbital of the radical with respect to the
normal axis of the membrane. In our studies, the correction term described by Gaffney[26] was used in the calculation of order parameters.
For the protein spin label (I_{103}) and in some cell systems for the
stearic acid spin label $(I_{1,14})$, the molecular motion was found to be
too fast for the above analysis. For these cases the data were analyzed
in terms of a correlation time (τ_c) for rotational motion.[27]

Measurement of Radical Reduction Kinetics

The plasma membrane of intact cells contains enzymatic activity
which causes the reduction of nitroxides to diamagnetic hydroxylamines.[28]
Since these enzymes are sulfhydryl-containing intrinsic proteins of the
plasma membrane, by studying the kinetics of the reduction under different experimental conditions, it is possible to extract valuable structural
information about the relationship of treatment to membrane structure.
In the reduction of the lipophilic labels we have excluded the possibility
that the decay process consists of reduction of the nitroxide by per-
meation of a water-soluble reducing agent into the lipid bilayer by
measuring the effect of 10^{-4} *M* ascorbic acid on the decay kinetics. Ascorbic acid was actually found to inhibit rather than enhance the decay
process. We have also excluded the possibility that the reduction process
is preceded by internalization of the lipophilic labels followed by cytoplasmic reduction of the nitroxide by observing quantitative regeneration
of the spin label spectrum. This was accomplished by briefly bathing
the intact cells in a solution of $K_4 Fe(CN)_6$ and remeasuring the electron
spin resonance signal. By this procedure we not only regenerate greater
than 98% of the original signal amplitude, but the characteristics of
the regenerated spectrum indicate that the spin label has remained in an
identical lipid environment throughout the time course of the experiments.

In all experiments of radical reduction kinetics the initial spin label
concentration was carefully adjusted so that pseudo first-order kinetics
were obeyed. In analyzing the kinetics of the enzymatic reduction of
the nitroxide spin labels we assumed that the reduction is irreversible
and that product inhibition is negligible. The reaction is monitored
throughout its total time course and a linear plot of spin concentration
(S) versus time (t) was constructed. The resulting data were then analyzed
using the integrated Michaelis-Menton equation.[28]

Measurement of Spin Exchange Processes

A further parameter investigated was the relationship between the
molar ratio of the spin label to phospholipid concentration and the spin

exchange rate per second (W_{ex}), for the control and treated cells. The rate of lateral diffusion is directly related to the number of encounters per second between randomly diffusing particles.

According to Trauble and Sackman,[29] the radical interaction in lipid lamellae containing a labeled lipid as one constituent may be determined by two factors: (1) the rate of lateral diffusion D of the labeled lipid molecules; and (2) the distribution of the lipids within the membrane, a factor that determines the average radical distance. These two mechanisms lead to completely different concentration dependencies of the exchange frequency W_{ex}.

Provided the radical interaction is determined by the lipid lateral diffusion, W_{ex} is proportional to the molar fraction (c/1 + c) of labeled lipid:

$$W_{ex} = \frac{4 \ d_c D_{diff}}{3\lambda F} \ \frac{c}{1 + c}$$

In this equation, d_c, λ and F are the critical radical interaction distance, the length of one diffusional jump and the area per lipid molecule, respectively. D_{diff} is the label lateral diffusion constant. This model is based on the assumption of a random lipid distribution within the lipid bilayers.

If W_{ex} is determined by the lipid distribution, that is by the formation of clusters of labeled lipid, these authors show that W_{ex} and c show a functional dependence according to

$$W_{ex} = \hat{W}_{ex} \ 1 - \pi \cdot d_c^2 \cdot n \sqrt{\frac{1 + c}{c}}$$

where n is the number of clusters/cm^2 and $\hat{W}_{ex}$ is the maximum exchange interaction in the interior of the cluster. By plotting W_{ex} both as a function of c/1 + c and as a function of $\sqrt{1 + c/c}$ respectively, one can distinguish between the above cases.

For the measurements of spin exchange processes the reductase activity of the plasma membrane was utilized to obtain several measurements of W_{ex} versus molar concentration of spin label. The spin label concentration was obtained by numerical integration of the electron spin resonance signal and comparison to a strong pitch standard run under identical conditions. Phospholipid concentrations were estimated from the protein concentration measured by the Lowry technique for each cell sample. It was assumed that in each case the phospholipid content was one-half the measured protein content. By assuming an average phospholipid molecular weight of 850, one could estimate the spin label phospholipid molar ratio. There are obvious quantitative inaccuracies in this procedure; however, since we were interested in comparing spin exchange processes between control and treated intact cells derived from either a cloned cell line or the same animal, we believe the procedure allows for such a comparison.

RESULTS

The Effect of SO_2/HSO_3^- on the Y-1 Adrenal Tumor Cell

It has been shown by radioactive tracer studies that inhalation of SO_2 leads to rapid accumulation of the gas or its hydrates into the systemic circulation to reach both the adrenal glands and other endocrine organs. It is thus pertinent to inquire at the molecular level the effect of this agent on the function of a selected endocrine cell system, the Y-1 adrenal tumor cell. The Y-1 adrenal tumor cell is stimulated to produce cAMP and steroid as a result of the peptide hormone ACTH binding to specific receptor sites on the surface of the plasma membrane. Results obtained in this laboratory have shown by the spin label technique that ACTH binding induces a conformational change in the receptor membrane-bound protein or glycoprotein molecules.[10] In view of the apparent affinity of SO_2 to protein molecules, it is possible that these peptide hormone responsive cells might be particularly sensitive to exposure to environmental agents such as SO_2 aq. To investigate this possibility, in a NaHSO$_3$ dose-response study we investigated the effect of SO_2/HSO_3^- on the ability of replicate (n = 3) Y-1 adrenal tumor cell cultures to produce cAMP and steroid both in the presence and absence of ACTH. In an additional study we examined the effect of 10^{-8} M NaHSO$_3$ on the kinetics of the 5-doxyl stearic acid spin label decay.

The Y-1 adrenal tumor cells were incubated in the presence of varying concentrations of NaHSO$_3$, pH 6.9, both in the presence and absence of 60 mU ACTH. Figure 1 shows that 90-minute incubations without ACTH, but in the presence of 10^{-3} - 10^{-9} M NaHSO$_3$, induced an increase in the mean extracellular accumulation of both steroids and cAMP. This nonspecific induction of the adenyl cyclase system with subsequent steroidogenesis was not related to sodium ions as incubation without ACTH, but in the presence of 10^{-3} - 10^{-9} M NaHSO$_4$ was without any effect. These findings appear to be related to the specificity of bisulfite/SO_2 for membrane-bound proteins.

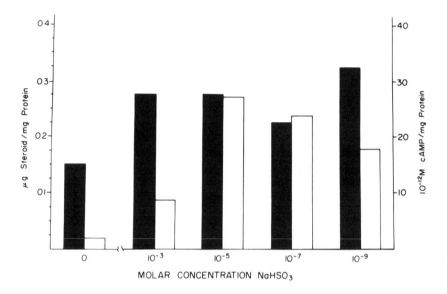

Figure 1. Effect of NaHSO₃ alone on the mean extracellular accumulation of steroid and cAMP.

To examine the effect of NaHSO$_3$ on ACTH-induced cAMP and steroid production, the Y-1 adrenal tumor cells were incubated with varying concentrations of NaHSO$_3$ in the presence of ACTH. The extracellular accumulation of both steroid (Figure 2) and cAMP (Figure 3) show a progressive increase with decreasing NaHSO$_3$ concentrations from 10^{-5} - 10^{-8} M. This potentiation of the ACTH response was observed following both 90- and 120-minute incubations. In a separate experiment, intracellular steroid production was also shown to give rise to a dose-response increase compared to control flasks; however, no apparent dose response was observed in intracellular cAMP in response to NaHSO$_3$ in the presence of ACTH. These studies suggest that NaHSO$_3$ alters both basal and ACTH-induced biological responses of the adrenal tumor cell by its ability to induce conformational changes in membrane-bound proteins, which activate the adenyl cyclase system with concomitant steroid production.

In Table 1(A), K$_m$ values for the reduction of the spin label 5-doxyl stearic acid are presented for 10-minute incubations with and without (50 mU/flask) ACTH both in the presence and absence of 10^{-8} M sodium bisulfite. Also included in the table are the extracellular cAMP and steroid levels measured after 90-minute incubations with cells of the same passage number as used for the spin label studies. As is evident from the table, almost identical K_m values are measured on incubation with ACTH with and without 10^{-8} M bisulfite and with 10^{-8} M bisulfite in the absence of ACTH. In each case a significant increase in K_m is observed over that measured for the control experiment. The data of Table I also illustrate that 10^{-8} M concentrations of bisulfite nonspecifically stimulate extracellular cAMP production to a level of 40% of that produced by the peptide hormone stimulation in the absence of bisulfite, whereas in the presence of 10^{-8} M bisulfite, extracellular levels are observed to be approximately 250% higher than those measured by ACTH stimulation. Nonspecific extracellular steroid production is also found to be approximately 50% of those measured by ACTH stimulation. In these experiments no significant increase in the extracellular levels of steroid was observed in the experiments in which both 10^{-8} M bisulfite and 50 mU ACTH were incubated with the cells compared to ACTH incubation alone.

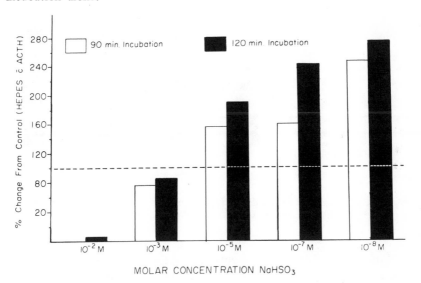

Figure 2. Effect of NaHSO$_3$ on the ACTH-induced mean extracellular accumulation of steroid compared to control (HEPES with ACTH, pH 6.9).

The Effect of NO_2, HSO_3^-/SO_2 and pH
on the Rat Alveolar Macrophage

In an attempt to separate effects on the plasma membrane of the macrophage, which might be pH related, from those that could be specifically attributed to the inhaled gases, experiments were performed to examine by the spin label technique both the effect of pH and the effect of both NO_2 and HSO_3^- at a constant pH of 7.4 on the physical properties of the macrophage plasma membrane. As in the case of the adrenal cell, the AM has nitroxide reductase activity, and the results to be described indicate specific effects of HSO_3^-/SO_2, NO_2 and pH on the alveolar macrophage plasma membrane intrinsic protein conformation and bulk phospholipid fluidity.

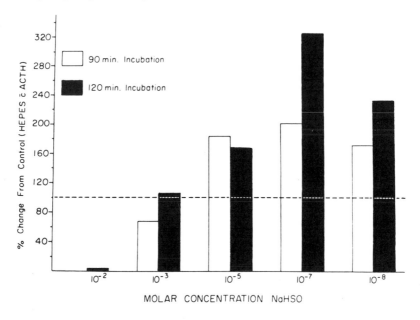

Figure 3. Effect of $NaHSO_3$ on the ACTH-induced mean extracellular accumulation of cAMP compared to control (HEPES with ACTH, pH 6.9).

Table II illustrates the data obtained by incubating freshly prepared Fisher rat AMs with logarithmic concentrations of $NaHSO_3$ and NO_2 in PBS, BSS at pH 7.4. For the case of the bisulfite experiments, dose-response data on bulk fluidity and nitroxide reduction kinetics were obtained for both 5-doxyl stearic acid and 16-doxyl stearic acid spin labels, whereas for the case of NO_2, experiments have so far been confined to the 5-doxyl stearic acid spin label. Macrophages obtained from several rats were pooled for each series of experiments so that the data obtained are internally consistent.

Table I. (A) Effect of $NaHSO_3$ in the Presence and Absence of ACTH,
on the km for the Decay of the 5-Doxyl Stearic Acid Spin-Labeled
Y-1 Adrenal Tumor Cell

Spin Label	Temperature ($^{\circ}$C)	Treatment	Time	Km
5-Doxyl Stearic Acid	37	Hepes	10'	$6.78 \times 10^{-3} M$
5-Doxyl Stearic Acid	37	Hepes + ACTH	10'	$1.6 \times 10^{-2} M$
5-Doxyl Stearic Acid	37	Hepes + $10^{-8} M$ $NaHSO_3$	10'	$2.7 \times 10^{-2} M$
5-Doxyl Stearic Acid	37	Hepes + $10^{-8} M$ $NaHSO_3$ + ACTH	10'	$1.74 \times 10^{-2} M$

(B) Effect of $NaHSO_3$, in the Presence and Absence of ACTH,
on the Y-1 Adrenal Tumor Cell Extracellular Accumulation
of cAMP and Steroid

Treatment	Temperature ($^{\circ}$C)	Time	Extracellular cAMP	Extracellular Steroid
Hepes	37	90'	$< 10 \times 10^{-12}$ M/Flask	0.6 μg/Flask
Hepes + $10^{-8} M$ $NaHSO_3$	37	90'	30×10^{-12} M/Flask	1.3 μg/Flask
Hepes + ACTH	37	90'	80×10^{-12} M/Flask	2.2μg/Flask
Hepes + $10^{-8} M$ $NaHSO_3$ + ACTH	37	90'	215×10^{-12} M/Flask	2.3 μg/Flask

Measurements of bulk fluidity observed using the spin label 5-doxyl stearic acid indicate that progressive increases in bulk fluidity are observed throughout the concentration range 10^{-8} M → 10^{-4} M $NaHSO_3$. No significant changes in K_m were observed for the nitroxide reduction kinetics at the 10^{-8} M concentrations, whereas a fivefold increase was observed at the 10^{-6} M concentration range and a twofold increase over the control values was observed at the 10^{-4} M concentration range. Although measurements of order parameter are considerably less precise with the spin label 16-doxyl stearic acid than with the 5-doxyl stearic acid, our measurements indicate that the dose-dependent increase in bulk fluidity observed with the latter is also observed for the 16-doxyl stearic acid experiments. A slight increase in measured K_m is observed at the 10^{-6} M bisulfite concentration and a more than threefold increase at the 10^{-4} M bisulfite concentration. In the experiments conducted with logarithmic concentrations of NO_2, an increase in bulk fluidity is measured at the 10^{-8} M

concentration with no further increases observed at the higher concentrations. A significant decrease in K_m was observed at the 10^{-8} M NO_2
concentration with further slight decreases observed at the higher concentrations. Also included in the table are calculated V_{max} values.

Table II. Effect of HSO_3^- and NO_2 on the Decay Kinetics
(Vmax and Km) and Order Parameter (S) for both the
5-Doxyl and 16-Doxyl Stearic Acid Spin-Labeled Alveolar Macrophages

Treatment	Spin Label	V_{max} (x 10^{-3} M/min)	K_m (x 10^{-3} M)	S_c
Control pH 7.4	5-Doxyl Stearic Acid	0.0058	0.0552	0.629
10^{-8} M HSO_3^-	5-Doxyl Stearic Acid	0.0059	0.0545	0.624
10^{-6} M HSO_3^-	5-Doxyl Stearic Acid	0.0249	0.2773	0.619
10^{-4} M HSO_3^-	5-Doxyl Stearic Acid	0.0120	0.1475	0.618
Control pH 7.4	16-Doxyl Stearic Acid	0.0031	0.0784	0.103
10^{-6} M HSO_3^- '	16-Doxyl Stearic Acid	0.0035	0.0938	0.100
10^{-4} M HSO_3^-	16-Doxyl Stearic Acid	0.0085	0.2867	0.096
Control pH 7.4	5-Doxyl Stearic Acid	0.0147	0.1173	0.645
10^{-8} M NO_2	5-Doxyl Stearic Acid	0.0069	0.0551	0.638
10^{-6} M NO_2	5-Doxyl Stearic Acid	0.0039	0.0288	0.638
10^{-4} M NO_2	5-Doxyl Stearic Acid	0.0035	0.0292	0.638

Measurements of order parameter reflect only the bulk fluidity of the
membrane lipids and do not determine whether there is a heterogeneous
distribution of membrane lipid with regions of high and low fluidity.
Galla and Sackmann[30] have shown that by measuring the spin exchange
frequency (W_{ex}) as a function of the molar percentage of spin label in a
model membrane system, a distinction could be made between a random
and a heterogeneous lipid distribution. For mixed lecithin-phosphatic acid

membranes they found that addition of both Ca^{2+} ions and polylysine leads to dramatic increases in the extent of phase separation in these membranes, which is the direct result of the formation of crystalline patches. Their results indicate that conformational changes in surface proteins may trigger in a cooperative way structural changes in the total lipid composition. It is clearly possible that conformational changes in intrinsic proteins may also give rise to similar phospholipid structural changes.

Measurements of spin exchange versus molar percent spin label for the 5-doxyl stearic acid spin label for control and 10^{-5} M concentrations of $NaHSO_3$ and NO_2 are illustrated in Figures 4 and 5, respectively. In each case a linear relationship exists between W_{ex} and $c/1 + c$, indicating that the spin exchange is controlled by lateral diffusion processes. Lateral diffusion constants calculated from these measurements are tabulated in Table III; however, it must be stressed that a comparison can only be made between calculated values for control and exposed values obtained within an experiment and not between experiments. The observed results in each case indicate that an increase in lateral diffusion constant is observed for exposed macrophages over that observed for the control cells. These results are consistent with the increases in bulk fluidity observed for the exposed cells (Table II). The effect of pH on bulk fluidity and nitroxide reduction kinetics throughout the pH range 5 → 8 is illustrated in Table IV, which shows that a gradual decrease in bulk fluidity is observed from pH 5-7 with a sharp decrease in bulk fluidity observed at pH = 8. Calculated K_m and V_{max} together with calculated pseudo first-order rate constant are also illustrated in the table. It can be seen that the variation in both K_m and V_{max} appears to be cyclical with the calculated pseudo first-order rate constant showing a progressive increase through pH range 5-8. Measurements of spin exchange versus molar percent spin label are illustrated in Figure 6. It can be seen that for both pH = 7 and pH = 8 a linear relationship exists between W_{ex} and $c/1 + c$.

Phospholipid Distributional Changes Between
Nontransformed and Chemically Transformed
Balb 3T3 Mouse Fibroblasts

The results of these ESR studies are tabulated in Table V. Slight differences in order parameter (S) were observed between both the nontransformed [Balb 3T3(1-13)] and transformed (Balb NQT) cells throughout the temperature range examined with the transformed cell line having the lower order parameter. These results are intermediate between order parameter values reported previously.[17,18] Measurements of order parameter reflect only the bulk fluidity of the membrane lipids and do not determine whether there is a heterogeneous distribution of membrane lipid with regions of high and low fluidity.

254

Therefore, additional studies of the dependence of W_{ex} on the molar percentage (c) of the spin label 5-doxyl stearic acid for both the non-transformed and transformed fibroblasts were conducted in these same cell clones (Figures 7 and 8). As illustrated for the nontransformed cell line (Figure 7), the spin exchange is governed by lateral diffusion processes ($W_{ex} \sim c/c + 1$). However, for the transformed cells (Figure 8), a much stronger radical interaction is observed. W_{ex} is now found to depend on $\sqrt{1 + c/c}$.

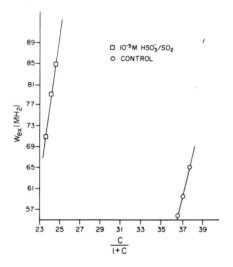

Figure 4. Effect of 10^{-5} M HSO_3^-/SO_2 on the relationship of electron spin resonance spin exchange (W_{ex}) to spin label phospholipid molar ratio (c/1 + c).

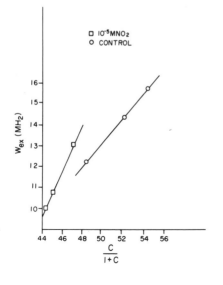

Figure 5. Effect of 10^{-5} M NO_2 on the relationship of electron spin resonance spin exchange (W_{ex}) to spin label phospholipid molar ratio (c/1 + c).

Table III. Calculated Lateral Diffusion Coefficients for
Alveolar Macrophages Following Various Treatments

Temperature ($^\circ$C)	Spin Label	Treatment	pH	D_{diff} (cm^2/sec)
24	5-Doxyl Stearic Acid	Control	7.4	1.0×10^{-7}
24	5-Doxyl Stearic Acid	$10^{-5}\ M\ NO_2$	7.4	1.95×10^{-7}
24	5-Doxyl Stearic Acid	Control	7.4	1.6×10^{-7}
24	5-Doxyl Stearic Acid	$10^{-5}\ M\ NaHSO_3$	7.4	2.4×10^{-7}
24	5-Doxyl Stearic Acid	–	8.0	5.2×10^{-8}
24	5-Doxyl Stearic Acid	–	7.0	2.5×10^{-7}

Table IV. Effect of pH on the Decay Kinetics (K_1, V_{max}, K_m) and
Order Parameter (S_c) for the 5-Doxyl Stearic Acid Spin-Labeled
Alveolar Macrophage

pH	Spin Label	k_1 (min^{-1})	V_{max} (x $10^{-3} M$/min)	K_m (x $10^{-3} M$)	S_c
5	5-Doxyl Stearic Acid	0.0811	0.037	0.457	0.621
6	5-Doxyl Stearic Acid	0.0907	0.162	1.781	0.625
7	5-Doxyl Stearic Acid	0.1102	0.063	0.574	0.629
8	5-Doxyl Stearic Acid	0.1276	0.277	2.170	0.650

DISCUSSION AND CONCLUSIONS

From our previous biophysical studies[10] we have concluded that
hormone-receptor interaction results in plasma membrane conformational
changes, which, in turn, cause increased lateral mobility of the resulting
hormone-receptor complexes. By the aggregation of several of the complexes, crystalline patches of phospholipid on the outer monolayer of the

plasma membrane are formed, which act as the transducer to activate the adenyl cyclase system that is bound to the inner surface of the plasma membrane.

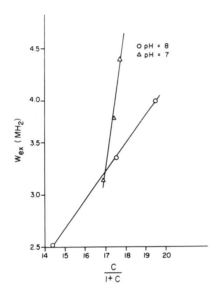

Figure 6. Effect of pH on the relationship of electron spin resonance spin exchange (W_{ex}) to spin label phospholipid molar ratio (c/1 + c)

Table V. Comparison of Temperature-Induced Changes in Order Parameter (S) Between 5-Doxyl Stearic Acid Spin-Labeled Untransformed (Balb 3T3 I-13) and Chemically Transformed (Balb NQT) Clones

Cell Type	Temperature (°C)	Order Parameter (S)
Balb 3T3 (I-13)	37	0.593
Balb 3T3 (I-13)	32	0.618
Balb 3T3 (I-13)	27	0.628
Balb NQT	37	0.587
Balb NQT	32	0.612
Balb NQT	27	0.623

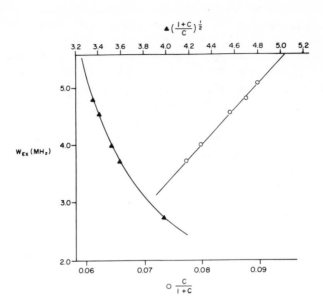

Figure 7. The relationship of electron spin resonance spin exchange (W_{ex}) to spin label phospholipid molar ratio (c/l + c) in the nontransformed cell (Balb 3T3).

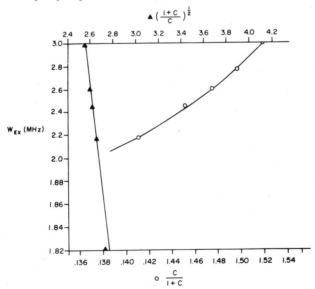

Figure 8. The relationship of electron spin resonance spin exchange (W_{ex}) to spin label phospholipid molar ratio (c/l + c) in the chemically transformed cell (Balb NQT).

The nonspecific stimulation of cAMP and extracellular steroid and the increase in K_m values for the reduction of the 5-doxyl stearic acid spin label observed for 10^{-8} M $NaHSO_3$ incubations with the Y-1 adrenal tumor cell compared to controls are consistent with the view that $NaHSO_3$ induces protein conformational changes that cause increased lateral diffusion of the receptor proteins. While the data provide no evidence of induced cluster formation, the observed effect is consistent with increased probability of receptor protein-adenyl cyclase interactions such as are observed in the increase in basal activity with increase in temperature.[31]

The ESR studies of the effects of NO_2 and bisulfite on the macrophage indicate that both NO_2 and bisulfite cause significant increases in bulk fluidity of the membrane phospholipids as evidenced by the measurements of order parameter. The effects of these two agents on nitroxide reduction kinetics is different, however. For both 5-doxyl stearic acid and 16-doxyl stearic acid increasing concentrations of bisulfite (10^{-6} M-10^{-4} M) lead to an increase in the calculated K_m and V_{max} and a decrease in the pseudo first-order rate constant over that found for controls. Since such behavior is not characteristic of a competitive inhibition, this possibility can be excluded. Thus, although it is well known that bisulfite anions covalently react with sulfhydryl groups, which the authors assume to be the reducing species associated with the nitroxide reduction, there appears to be no evidence of direct competition of the bisulfite for the reducing sites of the spin label. Whereas bisulfite caused an increase in the calculated K_m and V_{max} with increased concentration, increases in NO_2 concentrations between 10^{-8} M NO_2 and 10^{-6} M NO_2 lead to a decrease in both K_m and V_{max} and an apparent increase in the pseudo first-order rate constant. At a concentration of 10^{-4} M NO_2 the calculated K_m is increased over that observed at 10^{-6} M NO_2, but both the calculated V_{max} and pseudo first-order rate constant are found to decrease. Again this behavior is not characteristic of competitive inhibition by the NO_2. The observed behavior of both bisulfite and NO_2 towards the nitroxide reduction kinetics obeys the general relationships attributable to mixed inhibition. Considerable work will have to be undertaken in the future to fully understand the nature of the observed effects. However, these spin label studies indicate that even under conditions in which the buffering capacity of the lung is sufficient to maintain a fixed pH, both NO_2 aq and SO_2 aq at relatively low concentrations (10^{-6} M) can cause conformational changes to intrinsic membrane proteins of the alveolar macrophage. Studies of the effect of 10^{-5} M solutions of both NO_2 aq and HSO_3^- aq at pH 7.4 in each case show significant increases in the phospholipid lateral diffusion coefficients. It has been shown that hormone-receptor interaction induced protein conformational changes in the plasma membrane of the Y-1 adrenal tumor cells, which lead to significant changes in protein lateral diffusion coefficients in the absence of significant changes in phospholipid bulk fluidity.[10] For the case of the macrophage, the measured NO_2 aq and HSO_3^-/SO_2 aq induced increases in phospholipid bulk fluidity and phospholipid lateral diffusion coefficients together with induced protein conformational changes, suggesting that even at fixed pH these gases cause significant changes in protein lateral mobility.

The observed effects of increased pH on the nitroxide radical reduction kinetics are extremely complex. A progressive increase in pseudo first-order rate constant is observed throughout the pH range 5-8. However, the calculated values of V_{max} and K_m throughout this pH range are found to vary in a cyclical manner. The observed effect of increased pH on bulk fluidity is that a progressive decrease is observed from a pH of 5 through pH 7 and then an abrupt decrease is observed at pH 8. Measurements of phospholipid lateral diffusion coefficients for the AM between pH 7 and pH 8 are consistent with the observed bulk fluidity properties. The measured lateral diffusion coefficients show an order of magnitude decrease at pH 8 compared to that measured at pH 7. Based on the free volume model of protein diffusion, these results indicate that significant increases in lateral diffusion coefficients of membrane-bound proteins are also induced by an increase in the acidity of the macrophage environment.

The functional properties of the macrophage depend on membrane fluidity and intrinsic protein conformation and distribution. The biophysical studies described illustrate that both NO_2 aq and SO_2 aq cause easily detectable changes to the bulk fluidity and to protein conformation and mobility, independent of changes in the pH of the environment of the cell. However, changes in pH of the environment of the macrophage in the absence of these agents is also found to cause similar changes in the biophysical parameters measured. While it is difficult to relate the results of these *in vitro* biophysical studies to parameters obtained from *in vivo* exposure experiments, the biophysical studies suggest that processes that depend on bulk fluidity of the plasma membrane and correct membrane protein distribution might prove relative to alteration by these agents. One candidate membrane-associated enzyme receptor we have examined is the $Na^+K^+Mg^{2+}$ ATPase system of the alveolar macrophage, which has been suggested to act as a mechanoenzyme making phagocytosis possible through the conversion of chemical energy in the form of ATP to the mechanical energy required for attachment and ingestion. As illustrated in Table VI, *in vivo* exposure, both 10 ppm NO_2 for 4 hours and 5 ppm SO_2 for 2 hours significantly elevate AM ATPase activity of the Fisher rat. A further example of the ability of NO_2 to modify plasma membrane properties of the pulmonary macrophage was reported by Menzel (ACS Symposium on Environmental Health Effects, 1977), who found that exposures to NO_2

as short as two hours tend to increase the number and affinity of plasma-bound receptors for immunoglobulins and the plant lectins, Concanavallin A and wheat germ agglutinin. As pointed out by these authors, such results may stem from changes in the fluidity of the membrane or in the topographical distribution of the receptors. This hypothesis is fully consistent with the results of our biophysical studies.

Table VI. Effect of NO_2 and SO_2 Exposure on Alveolar Macrophage ATPase Activity

Group	N	Gas	Hours Exposure	ATPase μMoles P_i Liberated/ mg Protein/Hour)
Control	4	Room Air	4	5.6909 ± 0.4305
Exposed	4	10 ppm NO_2	4	6.7947 ± 1.5884
Control	6	Room Air	2	5.1 ± 0.5
Exposed	6	5 ppm SO_2	2	7.5 ± 0.04

Finally, the observation of phospholipid clusters in the chemically transformed Balb 3T3 cells, which are not observed in the nontransformed cells, strongly suggests that this might be the specific communication system that gives rise to the distinct physiological differences between the nontransformed and transformed cell lines. It might be suggested that in the case of the hormone-stimulated cells the effect is reversible, whereas the chemically transformed cells are locked into this stimulated configuration.

ACKNOWLEDGMENTS

This work was performed under Grants No. 5-SO7-RR05519-14 and No. 1-RO1-ES001162 from the National Institutes of Health.

REFERENCES

1. Goldstein, E., N. F. Peek, N. J. Parks, H. H. Hines, E. P. Steffey and B. Tarkington. *Am. Rev. Resp. Dis.* 115:403 (1977).
2. Rowlands, J. R., and E. M. Gause. *Arch. Intern. Med.* 128:94 (1971).
3. Rowlands, J. R., and E. M. Gause. Unpublished results.
4. Eickenroht, E. Y., E. M. Gause and J. R. Rowlands. *Environ. Lett.* 9:265 (1975).

5. Eickenroht, E. Y., E. M. Gause and J. R. Rowlands. *Environ. Lett.* 9:279 (1975).
6. Gause, E. M., N. D. Greene, M. L. Meltz and J. R. Rowlands. "Review of Multidisciplinary Studies in 'Biological Effects of Sulfur Oxides,' " paper presented at the Symposium on Biochemical Effects of Environmental Pollutants, Cincinnati, OH (1976).
7. Barry, D. H., and L. E. Mawdesley-Thomas. *Thorax* 25:612 (1970).
8. North, R. J. *J. Ultrastruct. Res.* 16:83 (1966).
9. Cross, C. E., M. G. Mustafa, P. Peterson and J. A. Hardie. *Arch. Intern. Med.* 127:1069 (1971).
10. Rowlands, J. R., and C. F. Allen-Rowlands. *Mol. Cell. Endocrinol.* 10:63 (1978).
11. Nicolson, G. L. *Int. Rev. Cytol.* 39:89 (1974).
12. Horwitz, A. F., M. E. Hatten and M. M. Burger. *Proc. Nat. Acad. Sci., U.S.* 71:3115 (1974).
13. Lehman, J. M., and J. R. Sheppard. *Virology* 49:339 (1972).
14. Kapeller, M., and F. Doljanski. *Nature (London) New Biol.* 235: 184 (1972).
15. Rosenblith, J. Z., T. E. Ukena, H. H. Yin, R. D. Berlin and M. J. Karnovsky. *Proc. Nat. Acad. Sci., U.S.* 70:1625 (1973).
16. Nicolson, G. L. *Nature (London) New Biol.* 233:244 (1971).
17. Barnett, R. E., L. T. Furcht and R. E. Scott. *Proc. Nat. Acad. Sci., U.S.* 71:1992 (1974).
18. Gaffney, B. J. *Proc. Nat. Acad. Sci. U.S.* 72:664 (1975).
19. Harper, J. E., and G. Brooker. *J. Cyc. Nucleotide Res.* 1:207 (1975).
20. Vernikos-Danellis, J., E. Anderson and L. Trigg. *Endocrinology* 79: 624 (1966).
21. Godin, Y., and S. L. Schrier. *Biochemistry* 9:4068 (1970).
22. Lowry, O. H., N. J. Rosebrough, A. L. Farr and R. J. Randall. *J. Biol. Chem.* 193:265 (1951).
23. Seelig, J. *J. Am. Chem. Soc.* 92:3881 (1970).
24. Hubbell, W. L., and H. M. McConnell. *J. Am. Chem. Soc.* 93:314 (1971).
25. McConnell, H. M., and M. McFarland (1970).
26. Gaffney, B. J. In: *Spin Labeling Theory and Applications*, Appendix IV, L. Berliner, Ed. (New York: Academic Press, 1976), p. 567.
27. Snipes, W., and A. Keith. *Res./Dev.* 21:22 (1970).
28. Segel, I. H. *Enzyme Kinetics* (New York: John Wiley & Sons, Inc., 1975), p. 46.
29. Trauble, H., and E. Sackmann. *J. Am. Chem. Soc.* 94:4499 (1972).
30. Galla, H. -J., and E. Sackmann. *Biochim. Biophys. Acta* 401:509 (1975).
31. Kreiner, P. W., J. J. Keirns and M. W. Bitensky. *Proc. Nat. Acad. Sci., U.S.* 70:1785 (1973).

CHAPTER 14

BIOMATHEMATICAL MODELING
APPLICATIONS IN THE EVALUATION
OF OZONE TOXICITY*

Frederick J. Miller
  Health Effects Research Laboratory
  Environmental Protection Agency
  Research Triangle Park, North Carolina 27711

INTRODUCTION

Environmental toxicologists are confronted with the formidable task of interpreting the manifold results from epidemiological, clinical, and animal studies on air pollutants such as ozone (O_3) and assessing their relevance and implications concerning pollutant levels to which man is exposed. Although the information obtained from epidemiological and clinical studies is directly related to the human experience, the fraction of the total toxicological data base represented by these types of studies is quite small in comparison to the data available from laboratory animal experiments. This data imbalance stems from the nature and inherent limitations of epidemiological and clinical studies. Uncertainties in epidemiological studies arise from the inability to control exposure levels, mobility of the population and the difficulty in accurately determining previous exposure history and important concomitant variables. The types of biological parameters that can be studied in most human clinical studies are usually limited to those for which noninvasive measurement techniques are available.

---

*This report has been reviewed by the Health Effects Research Laboratory, U.S. Environmental Protection Agency and approved for publication. Mention of trade names or commercial products does not constitute endorsement or recommendation for use.

By contrast to clinical studies, animal experimentation offers the researcher the choice of a wide range of concentrations, exposure regimens, chemical agents, biological parameters and animal species; however, this flexibility is not gained without expense. Obviously, relating dose response trends in animal studies to the human experience becomes more difficult. Furthermore, with a highly reactive gas such as O_3, the problem of estimating the effective dose delivered to the target organ is compounded. Species differences in the ratio of effective dose to exposure concentration must also be considered.

A unified approach, which involves nasal-pharyngeal removal experimental studies and lower airway mathematical modeling analyses,[1] has been formulated for evaluating the pulmonary toxicity of O_3. The applicability of this modeling approach can be illustrated by examining the similarity between man and laboratory animals in the regional pulmonary deposition of O_3. While determination of nasopharyngeal uptake of gaseous pollutants is straightforward, the tracheobronchial tree is, at present, relatively inaccessible to the local measurement of the uptake of highly reactive gases. Experimental estimates of nasopharyngeal removal determine necessary boundary conditions when modeling pollutant uptake in the lower airways. The technical inability of experimentally obtaining local lower airway O_3 uptake makes biomathematical modeling the method of choice for examining O_3 uptake in the lung. By presenting an overview of the major features of the model and by illustrating its application in the evaluation of O_3 toxicity, the intent of this chapter is to demonstrate the general approach for developing models for other gaseous pollutants.

NASOPHARYNGEAL REMOVAL OF OZONE

The degree to which environmental air pollutants can directly affect the respiratory surfaces of the lung is a function of the amount of pollutant and reaction products reaching the lower airways. The complex anatomical structure of the nose serves to humidify and regulate the temperature of inspired air, as well as to partially remove environmental contaminants. Nasal-pharyngeal removal serves to lessen the insult to the lung. Species differences in efficiency of nasopharyngeal removal must be considered when estimating O_3 concentrations responsible for observed pulmonary effects.

Aharonson et al.[2] demonstrated that the uptake coefficient, which defines the average flux of soluble vapor into the nasal mucosa per gas-phase unit partial pressure, increases with increasing airflow rate, provided the rate of uptake is proportional to the pressure of the vapor in the gas phase. Recently, Miller et al.[3] showed that O_3 satisfies this condition in guinea pigs and rabbits, and hence, the theoretical model approach of Aharonson et al.[2] can be used to study the effect of respiratory air flow-rate on removal of O_3 by the upper airways.

Removal of O_3 in the nasopharyngeal region of anesthetized rabbits and guinea pigs is markedly similar over a concentration range of 196-3920 $\mu g/m^3$ (0.1-2.0 ppm).[3] With a constant flowrate equal to the minute volume of the animal, the tracheal O_3 concentration was linearly related to the concentration of O_3 drawn through the isolated upper airways. Regression analyses showed that O_3 removal in the nasopharyngeal region is approximately 50% in both species. No difference in O_3 removal was observed between male and female rabbits.

By contrast to the 50% nasopharyngeal O_3 removal efficiency for guinea pigs and rabbits, estimates of nasopharyngeal removal of O_3 in dogs vary from 60-70%[4] to virtually 100% uptake.[5] Differences in the estimate of nasopharyngeal removal in dogs may be due to variations in the experimental technique used by these investigators because Yokoyama and Frank[6] found, when they repeated the procedure of Vaughn et al.[5] that some O_3 was lost due to absorption on the collection bag wall.

The findings of Yokoyama and Frank[6] on nasal versus oral uptake of O_3 are particularly important when modeling lower airway O_3 exposures during periods of exercise. These researchers found nasal uptake signifi-cantly exceeded (p < 0.01) oral uptake of O_3 in dogs when flowrates of 3.5 and 35 l/min were used. Since the route of breathing during exercise may be partially or completely oral, depending on the individual and the level of exercise, the tracheal O_3 concentration should be increased when modeling lower airway deposition in these instances.

The results of Moorman et al.[7] on the decomposition of O_3 in the nasopharynx of acutely versus chronically exposed beagle dogs illustrate another point relevant to determining boundary conditions in lower airway models. Dogs chronically exposed for 18 months to 1,960-5,880 $\mu g/m^3$ (1-3 ppm) using various daily exposure regimens were shown to have significantly higher mean tracheal concentrations of O_3 than animals tested after one day of exposure to the corresponding regimens. However, in view of the 25 $\mu g/m^3$ (0.0125 ppm) difference observed between animals chronically versus acutely exposed to 1,960 $\mu g/m^3$ (1 ppm), it is not likely that chronic exposure to more realistic environmental levels would result in tracheal O_3 concentrations significantly greater than those observed with acute exposure. Thus, it is probably reasonable to use the results of acute exposure studies on nasopharyngeal removal of O_3 to establish tracheal O_3 boundary concentrations in lower airway models.

LOWER AIRWAY OZONE UPTAKE MODEL

The airway luminal ozone concentration at a given level of the lung reflects the degree of pollutant insult at that level. Prior to striking the airway wall, ozone must first penetrate the mucous or surfactant layer lining the airway, depending on which region of the lung the ozone has reached. Chemical reactions may occur in these layers which increase the total ozone absorbed; however, the amount of ozone reaching the tissue may be reduced. Thus, the extent to which the model can incorporate chemical reactions is a function of how much is known about 1) the biochemical composition of the mucous and surfactant layers, and 2) the nature of the reactions of ozone with the components of these layers.

Only a brief overview of the major features of the lower airway model will be given, since a complete description of the model is given elsewhere.[8] The model characterizes the transport and removal of ozone within model segments that are partitions of individual airway generations and predicts the amount of ozone abosrbed in each generation of the tracheobronchial tree. The model formulation requires extensive use of lung morphometric data, such as are available for man,[9] rabbits, guinea pigs and rats.[10] Species anatomical properties and physiochemical properties are incorporated as model input parameters.

The effects of convection, axial diffusion and radial diffusion as well as the chemical reactions of ozone with various components of mucus are included in the model. A multistep procedure involving mass balance expressions in finite difference form is used to obtain solutions to the differential equations governing gas transport and removal. Respiration is assumed to be sinusoidal. To facilitate the numerical methods used to obtain solutions to the differential equations, the respiratory cycle was divided into a sequence of time steps.

Convection can be simulated by instantaneously moving a small gas volume through the airways during each time step. However, the sinusoidal flow pattern requires determining a different time increment for each step. The time increments can be found by integrating the gas volume flowrate curve over appropriately defined intervals and solving the resulting expressions for the necessary times. Since the gas volume increment was less than the model segment volume for model segments located towards the end of the conducting airways, some mixing occurred in each model segment in these regions during the convection phase. However, this situation was intentional because mixing does appear to occur in these regions.[11]

The model representation for axial diffusion considers the airways as an expanding cross-sectional channel. All pathways from the trachea to the alveoli are combined into one effective pathway whose cross-sectional area at any given distance from the trachea is equal to the summed cross-sectional area of all bronchial tubes at that distance. This approach has been used by a number of researchers.[12-16]

The effects on axial diffusion of both molecular diffusion and eddy diffusivity are considered by relating, in each airway generation at any given time, the diffusion coefficient to the mean axial gas velocity. Implementation of this model feature is based on a modification of the experimental results obtained by Scherer and co-workers.[17] These investigators used a five-generation glass tube model to empirically determine values for effective axial diffusivity for laminar flow of a gas in the conducting airways. The approach effectively replaces five generations by a single one with an average axial velocity and cross-sectional area equal to those of the initial generation, while retaining the effects of bifurcations, turbulence and secondary flows. Since predictions of ozone absorption in each generation of the tracheobronchial tree were desired, an average axial diffusion coefficient was computed by considering a given generation at each time step as the focal point of these five-generation simplifications. The desired result of increasing the diffusion coefficient was thereby achieved.

By contrast to the expanding cross-sectional area geometry used for axial diffusion, the appropriate geometry for radial diffusion is that of the lumen, which has a decreasing radius proceeding distally from the trachea. Radial diffusion is coupled with a scheme for incorporating chemical reactions to account for the removal of ozone from the lumen of an airway. As noted earlier, the biochemical composition of mucus, which lines the conducting airways, and surfactant, which lines the respiratory airways, indicates the manner in which chemical reactions should be modeled.

The amount of lipid, protein and carbohydrate per 100 g of human pulmonary secretions is given by Potter et al.[18] Also, the percent composition of neutral lipids and phospholipids, as well as the fatty acid composition of the major lipid components, has been characterized for human bronchial asthmatic mucus by Lewis.[19] Lewis concluded that the lipid composition of asthmatic mucus is typical of bronchial mucus because he found the same qualitative lipid composition in bronchitic mucus.

To date, detailed biochemical analysis has not been performed on the protein and carbohydrate components of bronchial mucus. The data of Levine et al.[20] on the amino acids found in human parotid glycoproteins were used to approximate the amino acid composition of glycoproteins in bronchial mucus. Of the amino acids present, only histidine is susceptible to oxidation by ozone.[21]

The amount of surfactant has been shown to correlate well with the amount of saturated lecithin in lung parenchyma and with alveolar surface area among various vertebrate species.[22] Saturated lecithin composes 90-95% of the recoverable lipid rabbit alveolar lavage fluid.[23,24] Thus, the respiratory regions of the lung are normally lined with saturated lecithin and are largely free of other lipids, proteins and carbohydrates.

Olefins are especially susceptible to oxidation by ozone. The reaction of ozone with olefins involves an initial reaction to form the zwitterion, and possibly ozonide, and then a reaction involving the peroxidation of the remaining olefins.[25] A direct attack of ozone upon the carbon-carbon double bonds of fatty acid is involved in the initiation of an irreversible reaction of ozone with olefins. Since the reactions of ozone with amines and olefins are diffusion controlled,[25,26] they can be characterized by an instantaneous reaction regime.

Basic to the instantaneous reaction scheme is the concept that the absorbed component, in this instance ozone, and the liquid-phase reactant cannot coexist in the same region of the liquid. Thus, the liquid can be considered as essentailly comprising two layers.[27] Diffusion of ozone occurs in the first layer, while diffusion of liquid-phase reactant takes place in the second layer. The airway luminal ozone concentration determines the location in the mucous layer of the reaction plane, the boundary between the two layers. As the luminal ozone concentration in a given conducting airway rises and falls during the breathing cycle, the reaction plane may be established anywhere from the gas-mucus interface to a position equivalent to ozone penetrating to the tissue.

Since surfactant, which lines the respiratory airways, consists almost entirely of saturated lecithin, the above formulation dicates that the mathematical model assumes no chemical reaction of ozone with the surfactant layer. However, since large amounts of unsaturated lipids are contained in lung tissue, it was assumed that ozone is instantaneously converted by chemical reaction when it reaches the tissue.

HUMAN LUNG MODEL RESULTS

Normal Respiration

While some of the results discussed in subsequent sections have been presented previously,[28] their inclusion here serves to illustrate the applicability of the model in evaluating ozone toxicity. Model estimates for the ozone dose delivered to the tissue in various generations of the human lung are presented in Figure 1 for tracheal ozone concentrations ranging from 62.5 to 4,000 $\mu g/m^3$ (0.03 to 2.04 ppm). Independent of the

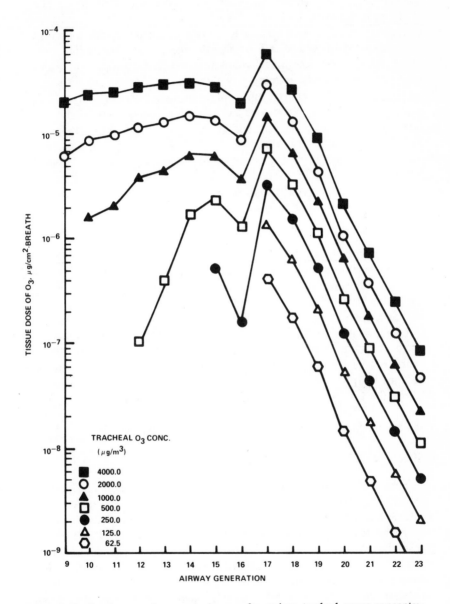

Figure 1. Predicted ozone dose curves in man for various tracheal ozone concentrations.[28] Model analyses used a tidal volume of 500 cc and a respiratory frequency of 15 breaths/min.

inhaled concentration, the first generation of respiratory bronchioles (generation 17) is predicted to receive the maximum ozone dose. The relationship between respiratory bronchiolar dose and the tracheal ozone concentration of inspired air can be shown to be linear with a positive slope on a logarithmic scale for tracheal concentrations exceeding 100 $\mu g/m^3$ (0.05 ppm). The model result that the first generation of respiratory bronchioles is the primary target during exposure to ozone is in agreement with the results from primate studies. Dungworth et al.[29] reported that the most obvious and consistent pulmonary lesions are found in respiratory bronchioles in primates exposed to ozone eight hours each day for seven days. With exposure to 392 $\mu g/m^3$ (0.2 ppm), the lesion was limited primarily to the proximal generation of respiratory bronchioles, while with exposure to 1568 $\mu g/m^3$ (0.8 ppm), the damage extended to occasionally involve proximal portions of alveolar ducts.

Penetration of ozone to conducting airway tissue (before generation 17) is not predicted for inhaled tracheal concentrations less than 125 $\mu g/m^3$ (0.06 ppm) (Figure 1). The concentration of mucous components, which will react with ozone, and the thickness of the mucous layer combine to deplete the absorbed ozone; however, increasing the tracheal concentration beyond 250 $\mu g/m^3$ (0.13 ppm) results in the conducting airway tissue deposition pattern becoming smoother, increasing in magnitude and including more airways.

Corresponding to the data of Figure 1, the relative contributions of mucus, conducting airway tissue and respiratory airway tissue to total ozone uptake are given in Table I. The percent uptake of ozone by mucus declines sharply as the inhaled tracheal concentration increases. Less than 8% of the total uptake is associated with conducting airway tissue. However, as the tracheal concentration of inspired ozone increases, respiratory airway uptake increases rapidly and plateaus at about 46.5%.

Effects of Tidal Volume on Dose

Tidal volume averages approximately 500 cc in adults during normal respiration. However, various levels of exercise result in increased tidal volume and respiratory frequency. Table II gives the predicted uptake of ozone by mucus and airway walls for various tidal volumes. There is a gradual decline in the percent uptake of ozone by mucus and conducting airway tissue as tidal volume increases from 500 to 1750 cc, while at the same time respiratory airway tissue uptake doubles.

While the increase in total uptake is mostly due to increased respiratory airway uptake, there is a plateau in the ozone dose to the first generation of respiratory bronchioles as tidal volume increases. This is evident from the ratios of the airway dose for various tidal volumes to the airway

Table I. Predicted Uptake of Ozone in the Human Lung by Mucus and Airway Walls for Various Tracheal Concentrations[a]

| | | Model Predicted % Uptake[b] | | |
| | | Airway Tissue | | |
Tracheal O_3 ($\mu g/m^3$)	Mucus (%)	Conducting (%)	Respiratory (%)	Total (%)
62.5	51.6	0.0	19.7	71.3
125.0	30.9	0.0	35.1	66.0
250.0	17.7	0.8	42.8	61.3
500.0	11.9	3.0	45.5	60.4
1000.0	9.9	5.2	46.4	61.5
2000.0	9.4	6.9	46.5	62.8
4000.0	9.2	7.9	46.6	63.7

[a]Tidal volume = 500 cc, respiratory frequency = 15 breaths/min.
[b]% uptake = 100 x total mass removed/total mass inhaled.

Table II. Predicted Uptake of Ozone in the Human Lung by Mucus and Airway Walls for Various Tidal Volumes[a]

| | | Model Predicted % Uptake[b] | | | |
| | | | Airway Tissue | | |
Tidal Volume (cc)	Inhale Time (sec)	Mucus (%)	Conducting (%)	Respiratory (%)	Total (%)
500	2.00	11.2	3.6	45.8	60.6
750	2.00	8.4	3.4	59.5	71.3
1,000	2.00	6.7	3.0	69.5	79.2
1,250	2.00	5.6	2.6	77.5	85.7
1,500	1.64	4.1	2.0	86.1	92.2
1,750	1.48	3.3	1.7	92.1	97.1

[a]600 $\mu g/m^3$ tracheal O_3 concentration.
[b]Uptake = 100 x total mass removed/total mass inhaled.

ozone dose during normal respiration shown in Figure 2. For tidal volumes greater than 1000 cc, there is a 2.5-fold increase in ozone dose to the first generation of respiratory bronchioles. However, ozone doses to the other generations of respiratory bronchioles (generations 18-19) continue to increase with increasing tidal volumes, and alveolar region (generations 20-23) doses increase dramatically. For example, with a tidal volume of 1750 cc,

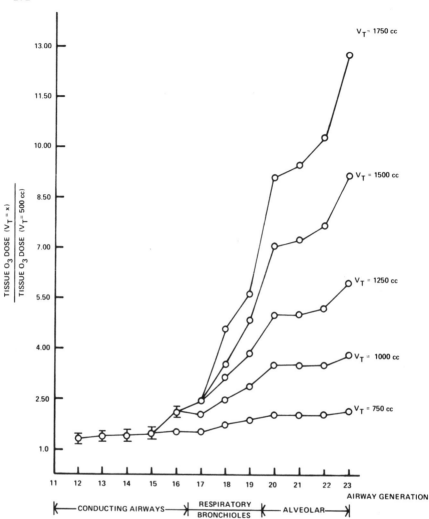

Figure 2. For various tidal volumes, the ratio of the ozone dose in a given airway generation to the dose during normal respiration in man (600 μg/m^3 tracheal concentration). Φ indicates that the dose ratios would all plot within this interval.

model predicted alveolar ozone doses are 9-13 times greater than during normal respiration.

Several investigators have reported significant decrements in various pulmonary function parameters in subjects who exercised while being exposed to ozone.[30-34] In the study of (Bates et al.[35]) two of the three subjects who exercised intermittently during ozone exposure showed changes in parameters of pulmonary function that were markedly accentuated over the levels observed in subjects who rested while being exposed to ozone. If nasopharyngeal removal of ozone in man should prove to be similar to that for rabbits and guinea pigs, the exposure level of 1470 $\mu g/m^3$ (0.75 ppm) used by Bates and co-workers would correspond to a tracheal ozone concentration of approximately 750 $\mu g/m^3$ (0.38 ppm).

With this concentration and a tidal volume of 1250 cc (sufficient to increase minute ventilation to the level Bates et al.[33] observed with exercise), a 2.4-fold increase in dose, compared to normal respiration, is predicted for the first generation of respiratory bronchioles. Also, a 5- to 6-fold incresae in alveolar region dose is predicted. Bates et al.[33] state that the measured fall in carbon monoxide extraction in one of the exercise subjects may indicate that this ozone concentration breathed during exercise has an effect at the alveolar level.

Although marked increases in alveolar region ozone doses are predicted with elevated tidal volumes, their magnitude is still considerably less than that predicted for the first generation of respiratory bronchioles (Table III). With a tidal volume of 500 cc, a tracheal ozone concentration of 750 $\mu g/m^3$ (0.38 ppm) results in the first generation of respiratory bronchioles (generation 17) receiving about 27 and 650 times the dose of alveolar generations 20 and 23, respectively. The corresponding dose ratios are about 13 and 261 for a tidal volume of 1250 cc.

The effects Bates et al.[33] observed indicating an increase in resistance in large and small airways may be reflected by the fact that ozone is predicted to penetrate to the tissue in a significant portion of the conducting airways (Table III). In subjects who intermittently exercised while being exposed to 725 $\mu g/m^3$ (0.37 ppm), Hazucha and co-workers[32] found significant alterations in pulmonary function, which they interpreted as being due to a decreased lung elastic recoil, increased airway resistance and small airway obstruction. If nasopharyngeal removal of ozone is again assumed to be approximately 50%, a tracheal ozone concentration of about 400 $\mu g/m^3$ (0.2 ppm) would correspond to the exposure level used by Hazucha et al.[32] From the dose ratios in Table III for 400 $\mu g/m^3$ (0.2 ppm), it can be seen that generations 14-16 are the only conducting airways for which the model predicts ozone penetration to the tissue. Thus, the model results support the statement[32] that "ozone, besides irritating larger airways and causing bronchoconstriction, may extend its action into terminal bronchioles and have possible indirect effect on lung parenchyma."

The variations in dose with changes in tracheal ozone concentration and tidal volumes indicate the need for establishing safe exposure levels

Table III. Ratio of Model-Predicted First Generation Human Respiratory
Bronchiolar Dose to Other Airway Doses for Various
Tidal Volumes and Tracheal Ozone Concentrations[a]

	Tracheal O_3 Concentration			
	400 $\mu g/m^3$		750 $\mu g/m^3$	
Airway	Tidal Volume		Tidal Volume	
Generation	500 cc	1250 cc	500 cc	1250 cc
10	-[b]	-	230.1	240.2
11	-	-	28.8	46.0
12	-	-	5.6	10.5
13	-	-	4.6	7.9
14	6.0	9.4	2.7	4.5
15	3.5	5.5	2.5	3.8
16	6.9	6.4	4.3	4.8
17	1.0	1.0	1.0	1.0
18	2.2	1.4	2.2	1.7
19	6.4	2.8	6.4	4.0
20	26.8	7.2	26.8	12.8
21	77.8	20.2	77.9	36.9
22	233.2	55.5	233.5	106.9
23	650.6	125.8	651.5	261.4

[a]Dose ratio for jth generation = tissue dose of 17th generation/tissue dose of jth generation, where the 17th generation is the first generation of respiratory bronchioles and
dose is expressed as $\mu g/cm^2$ - breath.
[b]Penetration of ozone to the tissue is not predicted by the model.

of this pollutant as a function of the physical activity of the exposed
population. The dose curves presented in Figure 3 further illllustrate this
point. If the level of exposure yields a tracheal ozone concentration of
200 $\mu g/m^3$ (0.1 ppm) and an individual's activity is such that his tidal
volume is 1750 cc, the predicted respiratory bronchiolar ozone dose
pattern is about the same as if they were exposed at rest to a tracheal
concentration of 750 $\mu g/m^3$ (0.38 ppm). Also, under these circumstances
alveolar region doses would be twice as great. The exercise dose curves
in Figure 3 are probably conservative because mouth breathing can be
associated with this level of activity and, therefore, the tracheal ozone
concentration is higher than that during normal respiration.

RABBIT AND GUINEA PIG LUNG MODEL RESULTS

Tissue ozone dose curves for rabbits and guinea pigs are shown in Figures 4 and 5, respectively. The available morphometric data for rabbits
and guinea pigs[10] is classified according to "morphometric zone" rather

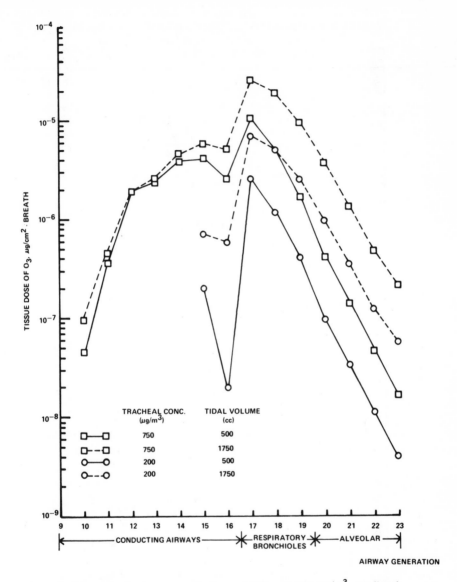

Figure 3. For tracheal ozone concentrations of 200 and 750 $\mu g/m^3$, predicted ozone dose curves for man during exercise (V_T = 1750 cc) vs normal respiration (V_T = 500 cc), where V_T = tidal volume.

than airway generation. Although several human airway generations may correspond to a morphometric zone in rabbits and guinea pigs, it is immediately apparent from Figures 1, 4 and 5 that a similarity exists

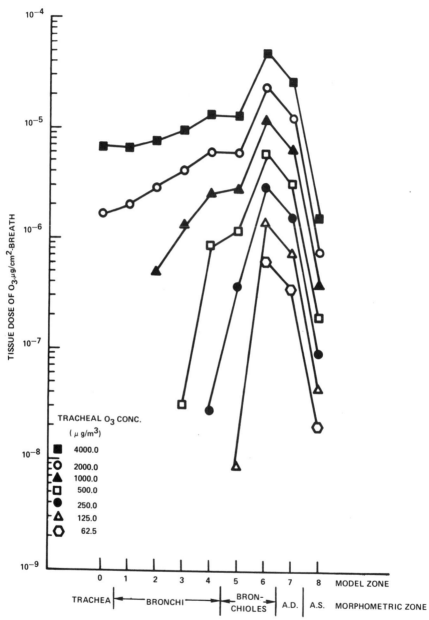

Figure 4. Predicted ozone dose curves in rabbits for various tracheal ozone concentrations.[28] Model analyses used a tidal volume of 21 cc and a respiratory frequency of 50 breaths/min.

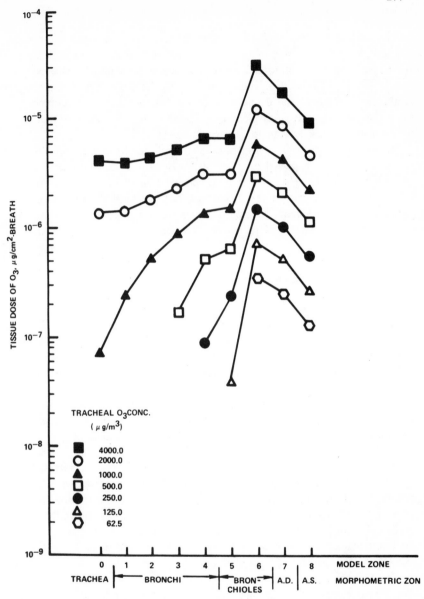

Figure 5. Predicted ozone dose curves in guinea pigs for various tracheal ozone concentrations.[28] Model analyses used a tidal volume of 1.8 cc and a respiratory frequency of 90 breaths/min.

between the three species in the general shape of the predicted dose curves. As for man, the model predicts that the respiratory bronchioles (model zone 6) of guinea pigs and rabbits receive the maximum ozone dose.

The decline in dose between alveolar ducts and sacs is much more pronounced in rabbits than in guinea pigs, which is probably a reflection of the larger proportion of total lung surface area represented by alveolar sacs in rabbits compared to guinea pigs. Thus, the somewhat closer resemblance to man of the ozone dose curves for rabbits is not surprising, since the surface areas of alveloar generations comprise a large proportion of the total surface area in the human lung.

At the lower tracheal concentrations studied, conducting airway doses are greater in guinea pigs than in rabbits. Also, for a tracheal ozone concentration of 1000 $\mu g/m^3$ (0.51 ppm), penetration of ozone to the tissue is predicted in the trachea of guinea pigs, but not until the second-order bronchi are reached in rabbits. With higher tracheal ozone concentrations, the conducting airway portions of the dose curves for both species are markedly similar.

INTERSPECIES DOSE CURVE COMPARISONS

A major goal of environmental toxicological studies on animals involves the eventual extrapolation of the results to man. This section illustrates how the model-predicted dose curves can be used to estimate comparable exposure levels for guinea pigs, rabbits and man and account for species differences in the ratio of effective dose to exposure concentration.

Figure 6 illustrates the differences between guinea pigs, rabbits and man in respiratory bronchiolar ozone dose for various tracheal ozone concentrations. Compared to man, rabbits and guinea pigs exhibit a more gradual decline in respiratory bronchiolar ozone dose as the tracheal ozone concentration decreases from 100 $\mu g/m^3$ (0.05 ppm). For tracheal ozone concentrations greater than 100 $\mu g/m^3$ (0.05 ppm), regression analysis applied to the data in Figure 6 shows that the relationship between respiratory bronchiolar dose and the inhaled tracheal concentration is linear on a logarithmic scale for all three species. Analysis shows the predicted respiratory bronchiolar dose for rabbits is 80% of that for man and twice that for guinea pigs for any given tracheal ozone level exceeding 100 $\mu g/m^3$ (0.05 ppm).

To illustrate how Figure 6 may be used to compare toxicological data from experiments on guinea pigs and rabbits and to predict the exposure level equivalent in man, consider the following example. Suppose an experiment has been conducted in which guinea pigs were exposed to 2000 $\mu g/m^3$ (1.02 ppm). Allowing for approximately 50% nasopharyngeal removal,[3] this exposure level corresponds to a tracheal ozone concentration of 1000 $\mu g/m^3$ (0.51 ppm). From Figure 6 it can be seen that a 1000 $\mu g/m^3$ (0.51 ppm) tracheal ozone concentration in guinea pigs

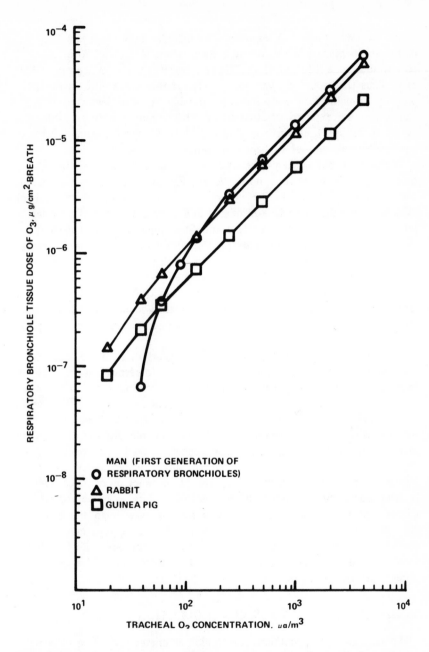

Figure 6. Respiratory bronchiolar ozone dose vs concentration of ozone in the trachea of rabbits, guinea pigs and man.[28] Model analyses for man used a tidal volume (V_T) of 500 cc and a respiratory frequency (f) of 15 breaths/min. For rabbits, V_T = 21 cc and f = 50 breaths/min, while for guinea pigs, V_T = 1.8 cc and f = 90 breaths/min.

yields the same predicted respiratory bronchiolar dose as does a 500 $\mu g/m^3$ (0.26 ppm) tracheal ozone concentration in rabbits. Also, this corresponds to a 420 $\mu g/m^3$ (0.21 ppm) tracheal ozone concentration in man. Thus the model predicts that, as far as the ozone dose to the respiratory bronchioles is concerned, comparable exposure levels occur if guinea pigs are exposed to 2000 $\mu g/m^3$ (1.02 ppm), rabbits to 1000 $\mu g/m^3$ (0.51 ppm) and man to 840 $\mu g/m^3$ (0.42 ppm), assuming 50% nasopharyngeal removal of ozone in man.

If the same biological parmeters have not been measured in these species at dose-equivalent exposure levels, Figure 6 can be used to design new studies to fill gaps in the current data base. This will allow mechanism of action hypotheses to be formulated and tested. Utilizing the model-predicted dose curves (Figures 1, 4 and 5), comparisons of equivalent exposure levels can also be made for the conducting airways and alveolar regions. However, the variability between species in the dose curve patterns for these regions probably does not justify such comparisons at this time. These comparisons should await additional data on mucus composition in rabbits and guinea pigs, diffusion rates for ozone in mucus and surfactant, thickness of the mucous layer in guinea pigs and man, mucus production rates, etc.

Illustrative of other effective dose comparisons that can be made concern estimating the total μg of ozone delivered to respiratory bronchiolar tissue for any fixed length of exposure. After accounting for differences between rabbits, guinea pigs and man in respiratory frequency, the respiratory bronchioles of rabbits recieve 3.2 times more μg of ozone than do the first generation of respiratory bronchioles in man for the same length of exposure to a given tracheal ozone concentration greater than 100 $\mu g/m^3$ (0.10 ppm). Compared to man, the analogous estimate for guinea pigs is 2.4 times more μg of ozone. Also, 1.33 times more μg of ozone is delivered to the respiratory bronchioles of rabbits, as comapred to guinea pigs, for the same length of exposure. Between man and laboratory animals, effective dose on a μg ozone/cm^2-breath basis may relate the best for comparison of pulmonary function parameter effects, while total μg of ozone may be more relevant for comparison of lesions and other structural changes.

MODEL SENSITIVITY TO MUCUS PRODUCTION

To implement the mathematical model, estimates had to be obtained for K, the effective production rate (at the airway wall) of the components of mucus with which ozone will react.[1] As the morphometric data for rabbits and guinea pigs[10] are only available by "zone," while the data for man are given by dichotomously branched generation,[9] "mucus production zones" were formed for man by considering which generations had drained the same percentage of mucus-lined airway surface area as had drained a given morphometric zone in rabbits and guinea pigs. K was assumed to decrease linearly with respect to mucous zone from a maximum value at the trachea to zero at the respiratory bronchioles. Obtaining estimates of K involved $\overline{K}_O$, the average rate of production per unit area of reaction components. The question arises of the sensitivity of the model-predicted respiratory bronchiolar doses to the value of $\overline{K}_O$.

The predicted respiratory bronchiolar dose in guinea pigs, rabbits and man as a function of $\overline{K}_O$ is shown in Figures 7 and 8 for tracheal ozone concentrations of 325 $\mu g/m^3$ (0.17 ppm) and 750 $\mu g/m^3$ (0.38 ppm), respectively. If $\overline{K}_O$ (standard) overestimates the correct value, there is no difference in the predicted respiratory bronchiolar dose for any of these species at either concentration. However, the interpretation of the model results is not as clear, if $\overline{K}_O$ (standard) underestimates the correct value.

In rabbits and guinea pigs, only if the correct value has been underestimated at least 25-fold at the lower concentration (Figure 7) and at least 50-fold at the higher concentration (Figure 8) does the model predict significantly different respiratory bronchiolar ozone doses. Even then, the general shape of the dose curve is still the same for the respiratory regions. Only the relative location on the dose-axis and the shape of the conducting airways portion of the curve are changed.

Predicted respiratory bronchiolar doses in man are more sensitive to the value of $\overline{K}_O$ than are the doses in guinea pigs and rabbits. The sensitivity is again a function of the tracheal ozone level with significant differences in respiratory bronchiolar dose being associated with 10-fold and 25-fold underestimation of the correct value of $\overline{K}_O$ for 325 $\mu g/m^3$ (0.38 ppm), respectively.

In the conducting airways, decreasing $\overline{K}_O$ allows penetration of ozone to the tissue higher in the respiratory tract. However, the magnitudes of the resulting doses for all three species still do not approach those predicted for the respiratory bronchioles. On the other hand, increasing $\overline{K}_O$ results in less ozone penetrating to conducting airway tissue.

Although not presented, a 25% increase in the thickness of the mucous layer decreases conducting airway tissue doses but has no effect on respiratory region ozone doses. Shifts in the earliest region of the conducting airways predicted to have penetration of ozone to the tissue can occur, especially with lower tracheal ozone concentrations.

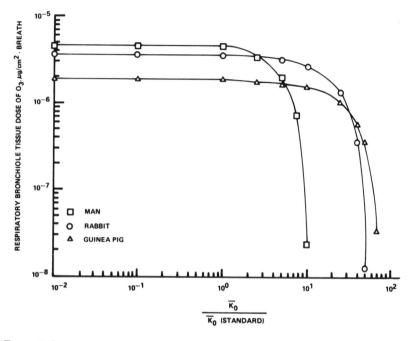

Figure 7. Sensitivity of respiratory bronchiolar ozone dose to the average current of mucous reaction components (325 μg/m^3 tracheal concentration). See Figure 6 for the tidal volume and respiratory frequency used for each species.

AREAS OF STUDY RESULTING IN REFINEMENT OF BIOMATHEMATICAL MODELS

While the model analyses have used the best available biological and physiochemical data, there are many research areas that would provide better data for incorporation into lower airway models. Moreover, a substantial portion of these data are not specific to a lower airway model for ozone uptake and would have general applications in modeling any gaseous pollutant.

The work of Scherer et al.[17] on the effective axial diffusivity in the lungs of man as a function of the mean axial gas velocity needs to be extended to guinea pigs and rabbits. The vorcities imparted by the bifurcation angles and branching patterns in rabbits and guinea pigs may result in significantly altered diffusion coefficients in the conducting airways. This could change the dose curves for these regions but it is not likely to influence respiratory dose patterns, since molecular diffusion is the dominant mechanism facilitating gas exchange in the respiratory region.

The data of Luchtel[35] on the thickness of the mucous layer in rabbits have been used as the basis for assigning thickness values in guinea pigs

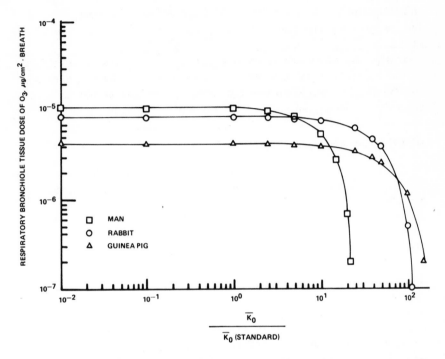

Figure 8. Sensitivity of respiratory bronchiolar ozone dose to the average current of mucous reaction components (750 $\mu g/m^3$ tracheal concentration). See Figure 6 for the tidal volume and respiratory frequency used for each species.

and man. Studies on mucus thickness in guinea pigs and man are needed before more definitive comparisons of conducting airway ozone doses can be made between these species.

Also, rates of production of mucus in various parts of the conducting airways need to be established. The current model assumed that the average rate of production of mucous reaction components is the same for rabbits, guinea pigs and man. Additional experimental data are needed to judge the validity of this assumption.

The diffusion coefficient of ozone in mucus and surfactant needs to be established. The diffusion coefficient of ozone in mucus is probably less than that for ozone in water. However, ciliary activity effectively increases the diffusion coefficient. Thus, the value of the diffusion coefficient for ozone in water, which was used in the model analyses, may be close to the correct value. Since the diffusion coefficient is important in establishing the mass flowrate of ozone from the lumen into the mucus and surfactant layer, better estimates of the diffusion coefficients would allow a more accurate prediction of absolute differences in dose between species.

Lastly, nasopharyngeal removal of ozone in man needs to be established for different flowrates, as well as for elevated flowrates in guinea pigs and rabbits. This would provide the correct tracheal ozone boundary conditions for modeling lower airway deposition of ozone during exercise.

DISCUSSION AND CONCLUSIONS

It has been the intent of this chapter to illustrate that the combination of upper airway empirical studies and lower airway mathematical models provide a unified approach to evaluating the toxicity of ozone and to demonstrate a general approach for developing models for other gaseous pollutants. The environmental toxicologist must make judgements concerning the validity of extrapolating to man the results obtained from animal studies. Differences between man and experimental animals in the effective dose delivered to the target organ must be considered. To achieve these objectives, greater emphasis must be placed on developing mathematical models for pollutant target organs. The consistency and similarity of the model-predicted ozone dose curves for rabbits, guinea pigs and man lend strong support to the validity of extrapolating to man the results obtained on animals exposed to ozone.

REFERENCES

1. Miller, F. J. "A Mathematical Model of Transport and Removal of Ozone in Mammalian Lungs," Ph.D. Thesis, North Carolina State University, Raleigh, NC.
2. Aharonson, E. F., H. Menkes, G. Gurtner, D. L. Swift and D. F. Proctor. "Effect of Respiratory Airflow Rate on Removal of Soluble Vapors by the Nose," *J. Appl. Physiol.* 37:654-657 (1974).
3. Miller, F. J., C. A. McNeal, D. E. Gardner, D. L. Coffin and D. B. Menzel. "Nasopharyngeal Removal of Ozone in Rabbits, Guinea Pigs, and Rats," *J. Appl. Physiol.* Submitted for publication.
4. Frank, N. R., E. Yokayama, S. Watanabe, D. E. Sherry and J. D. Brain. "An Ozone Journal," paper presented at the AMA Air Pollution Research Conference, New Orleans, October 7, 1970.
5. Vaughan, T. R., Jr., L. F. Jennelle and T. R. Lewis. "Long-Term Exposure to Low Levels of Air Pollution; Effects on Pulmonary Function in the Beagle," *Arch. Environ. Health* 19:45-50 (1969).
6. Yokoyama, E., and R. Frank. "Respiratory Uptake of Ozone in Dogs," *Arch. Environ. Health* 25:132-138 (1972).
7. Moorman, W. J., J. J. Chmiel, J. F. Stara and T. R. Lewis. "Comparative Decomposition of Ozone in the Nasopharynx of Beagles," *Arch. Environ. Health* 26:163-155 (1973).

8. Miller, F. J., D. B. Menzel, D. L. Coffin, R. J. Monroe and H. L. Lucas. "A Mathematical Model for Pulmonary Deposition of Ozone," *Respir. Physiol.* Submitted for publication.

9. Weibel, E. R. *Morphometry of the Human Lung* (New York: Academic Press, Inc., 1963).

10. Kiment, V. "Similarity and Dimensional Analysis, Evaluation of Aerosol Deposition in the Lungs of Laboratory Animals and Man," *Folia Morphol.* 21:59-64 (1973).

11. Altshuler, B., E. D. Palines, L. Yarmus and N. Nelson. "Intrapulmonary Mixing of Gases Studied with Aerosols," *J. Appl Physiol.* 14:321-327 (1959).

12. Davidson, M. R., and J. M. Fitz-Gerald. "Transport of O_2 Along a Model Pathway Through the Respiratory Region of the Lung," *Bull. Math. Biol.* 36:275-303 (1974).

13. Paiva, M. "Gas Transport in the Human Lung," *J. Appl. Physiol.* 35:401-410 (1973).

14. Pedley, T. J. "A Theory for Gas Mixing in a Simple Model of the Lung," in *Fluid Dynamics of Blood Circulation and Respiratory Flow*, AGARD Conference Proceedings No. 65 (1970).

15. Scherer, P. W., L. H. Shendalman and N. M. Greene. "Simultaneous Diffusion and Convection in a Single Breath Lung Washout," *Bull. Math. Biophys.* 34:393-412 (1972).

16. Yu, C. P. "On Equations of Gas Transport in the Lung," *Resp. Physiol.* 23:257-266 (1975).

17. Scherer, P. W., L. H. Shendalman, N. M. Greene and A. Bouhuys. "Measurement of Axial Diffusivities in a Model of the Bronchial Airways," *J. Appl. Physiol.* 38:719-723 (1975).

18. Potter, J. L., L. W. Matthews, J. Lemm and S. Spector. "Human Pulmonary Secretions in Health and Disease," *Ann. N.Y. Acad. Sci.* 106:692-697 (1963).

19. Lewis, R. W. "Lipid Composition of Human Bronchial Mucus," *Lipids* 6:859-861 (1971).

20. Levine, M. J., J. C. Weill and S. A. Ellison. "The Isolation and Analysis of a Glycoprotein from Parotid Saliva," *Biochim Biophys. Acta* 188:165-167 (1969).

21. Mudd, J. B., R. Leavitt, A. Ongun and T. T. McManus. "Reaction of Ozone with Amino Acids and Proteins," *Atmos. Environ.* 3:669-682 (1969).

22. Clements, J. A., J. Nellenbogen and H. J. Trahan. "Pulmonary Surfactant and Evolution of the Lungs," *Science* 169:603-604 (1970).

23. Balis, J. U., S. A. Shelley, M. J. McCue and E. S. Rappaport. "Mechanisms of Damage to the Lung Surfactant System—Ultrastructure and Quantitation of Normal and *in vitro* Inactivated Lung Surfactant," *Exp. Mol. Pathol.* 14:243-262 (1971).

24. Hurst, D. J., K. H. Kilburn and W. S. Lynn. "Isolation and Surface Activity of Soluble Alveolar Components," *Resp. Physio.* 17:72-80 (1973).

25. Menzel, D. B. "The Role of Free Radicals in the Toxicity of Air Pollutants (Nitrogen Oxides and Ozone)," in *Free Radicals in Biology*, Vol. II (New York: Academic Press, 1976), pp. 181-202.

26. Criegee, R. In: *Peroxide Reaction Mechanisms*, J. D. Edwards, Ed. (New York: Wiley-Interscience, 1962), pp. 29-39.
27. Astarita, G. *Mass Transfer with Chemical Reaction* (New York: Elsevier Publishing Co., 1967), p. 187.
28. Miller, F. J., D. B. Menzel and D. L. Coffin. "Similarity Between Man and Laboratory Animals in Regional Pulmonary Deposition of Ozone," *Environ. Res.* Submitted for publication.
29. Dungworth, D. L., W. L. Castleman, C. K. Chow, P. W. Mellick, M. G. Mustafa, B. Tarkington and W. S. Tyler. "Effect of Ambient Levels of Ozone on Monkeys," *Fed. Proc.* 34:1670-1674 (1975).
30. Kagawa, J., and T. Toyama. "Effects of Ozone and Brief Exercise on Specific Airway Conductance in Man," *Arch. Environ. Health* 30:36-39 (1975).
31. Hackney, J. D., W. S. Linn, D. C. Law, S. K. Karuza, H. Greenberg, R. D. Buckley and E. E. Pedersen. "Experimental Studies on Human Health Effects of Air Pollutants. III. Two-Hour Exposure to Ozone Alone and in Combination with Other Pollutant Gases," *Arch. Environ. Health* 30:385-390 (1975).
32. Hazucha, M., F. Silverman, C. Parent, S. Field and D. V. Bates. "Pulmonary Function in Man After Short-Term Exposure to Ozone," *Arch. Environ. Health* 27:183-188 (1973).
33. Bates, D. V., G. M. Bell, C. D. Burnham. M. Hazucha, J. Mantha, L. D. Pengelly and F. Silverman. "Short-Term Effects of Ozone on the Lung," *J. Appl. Physiol.* 32:176-181 (1972).
34. Folinsbee, L. J., F. Silverman and R. J. Shephard. "Exercise Response Following Ozone Exposure," *J. Appl. Physiol.* 38:996-1001 (1975).
35. Luchtel, D. L. "Ultrastructural Observations on the Mucous Layer in Pulmonary Airways," ICCB Abstract No. 1048, *J. Cell. Biol.* 70:350a (1976).

CHAPTER 15

INFLUENCE OF ATMOSPHERIC PARTICULATES ON PULMONARY ABSORPTION PHENOMENON*

J. M. Charles

Environmental Protection Agency
Health Effects Research Laboratory
Research Triangle Park, North Carolina 27711

D. B. Menzel

Departments of Pharmacology and Medicine
Duke University Medical Center
Durham, North Carolina 27710

INTRODUCTION

Man is exposed to large numbers of aerosols that may contain toxicants. Liquid sprays are the principal means of application of herbicides and pesticides. Liquid aerosols of respirable diameter are likely to result during spray applications, especially with concentrated or low-volume sprayers. Although considerable effort has been expended to reduce the amount of spray generated as a respirable aerosol, complete removal of respirable aerosols does not appear to be technically feasible. Home and garden applications of various compounds are also likely to produce respirable aerosols. During the production and compounding of industrial compounds, workers are likely to be exposed to both vapors and aerosols. Inhalation, which represents a potentially important route of absorption of these toxicants, has been little explored. From what has been found, inhalation appears to represent a highly efficient means of absorption of

---

*Mention of trade or commercial products does not constitute endorsement or recommendation for use by the U.S. Environmental Protection Agency.

toxicants and drugs. Little is known regarding the mechanisms of direct absorption of drugs and toxicants from the lung, independent of phagocytes and mucociliary clearance.

The respiratory tract epithelium has been shown to behave as a highly porous membrane permeable to a number of solutes. Removal of organic compounds from the airway has been studied by Enna and Schanker[1] by administering intratracheally a small volume of Krebs Ringer phosphate solution containing ^{14}C-labeled compounds to anesthetized rats. The absorption of compounds from the airways following intratracheal instillation represents an absorption route analogous to that of inhalation of a liquid aerosol. One group of compounds he studied was the lipid-insoluble neutral compounds, including urea, erythritol, mannitol and sucrose. When administered over a 100 to 1,000-fold range of concentration, they disappeared from the lungs at rates directly proportional to the concentration.[1] The relative rates of absorption ranked in the same order as the diffusion coefficients of the compounds. Simple diffusion appears to account for removal. Because of the extremely low lipid solubility of these compounds, Enna and Schanker suggested that absorption was predominantly by passage through aqueous channels or pores in the membrane rather than through lipid regions. Diffusion through at least three different populations of pore size could explain the absorption of these compounds. The smallest-diameter pore prevents the diffusion of all the saccharides relative to that of urea. A larger diameter pore allows the diffusion of only erythritol, while the largest pore size permits the passage of all compounds except dextran (molecular weight-70,000).

In another study,[2] involving organic anions and cations, sulfanilic acid, tetraethyl ammonium ion, p-aminohippuric acid, p-acetylaminohippuric acid and procainamide ethobromide appeared to be absorbed by diffusion through aqueous pores since their absorption was nonsaturable and roughly related to molecular size rather than to partition coefficient.[2]

Phenol red, a lipid insoluble organic anion, is an exception and is absorbed from the rat lung not only by diffusion but also in part by a carrier-type transport process ($t_{1/2}$ = 20 minutes). The carrier-mediated process, which becomes saturated at high concentrations of the dye, is inhibited by certain organic anions including benzylpenicillin and cephalothin.[3]

The main barrier to the diffusion of water-soluble compounds is the alveolar membrane. In studies with the isolated perfused dog lung, Taylor and Gaar[4] calculated an equivalent pore radius of 8-10 Å for the alveolar membrane and a much larger radius for the capillary endothelium.

Lipid-soluble compounds are thought to be absorbed mainly by diffusing through lipoid regions of the membrane. Burton and Schanker[5] administered five antibiotics intratracheally to anesthetized rats. The $t_{1/2}$ ranged from 1.9-33 minutes. Chloramphenicol was absorbed most rapidly followed by doxycycline, erythromycin and tetracycline, with benzyl penicillin showing the slowest rate. A comparison of pulmonary absorption rate, molecular weight and chloroform/water partition coefficient of the drugs indicated that lipid solubility was more closely associated with the relative rate of absorption than molecular size. Corticosteroids are rapidly absorbed from the respiratory tract with $t_{1/2}$ of absorption ranging from 1.0-1.7 minutes.[6]

In other studies on lipid soluble compounds, Normand et al.[7] investigated the permeability of alveoli and capillaries in the fluid-filled lungs of the fetal lamb. They reported that urea, thiourea and N-ethylthiourea penetrated the alveolar wall at rates that increased with the lipid solubility of the compounds. Earlier, Taylor et al.[8] in a study of the alveolar membrane of the isolated perfused dog lung, reported that the permeability coefficient of a lipid-soluble compound, dinitrophenol, was much greater than that of lipid-insoluble compounds, such as glucose and urea.

The actual site of solute absorption is unknown. Absorption could occur across the alveolar epithelium, across the broncheolar epithelium of the airways or at a combination of these sites. Burton and Schanker[9] have reported that lung absorption rates are increased by at least twofold when solutes are administered as aerosols.

PULMONARY ABSORPTION OF RESPIRABLE TOXICANTS

Methods

Two experimental techniques were employed to study the pulmonary absorption of two distinctly different types of toxicants, sulfate salts and herbicides. The isolated, ventilated and perfused rat lung was prepared by a modification of the technique of Niemeier and Bingham.[10] Tyrode's solution, pH 7.35, 37°C, was modified by the addition of 35.0 g/l PVP (polyvinylpyrrolidone), average molecular weight 40,000, to maintain a physiological oncotic pressure. Perfusion of the lungs was via the pulmonary artery. The perfusion rate was maintained at a constant value of 2.0 ± 0.1 ml/min. The lungs were suspended by the tracheal cannula in an artificial thorax at 25°C where respiration was maintained mechanically by an alternating negative pressure of -3 to -15 cm of water. The venous effluent from the lung could be collected directly in a fraction collector using a peristaltic pump.

Kinetic characterization of airway absorption by the pulmonary circulation was accomplished by intratracheal injection of the labeled compound in 100 μl of isotonic sucrose at pH 7.4.

To determine the reproducibility of intratracheal instillation as a route of exposure of the lung to sulfate salts, a single injection of 100 μl of isotonic sucrose containing 4 μCi of $[^{35}S]$-Na_2SO_4 was given, and the perfusion and ventilation were continued for five minutes. The lung was then removed from the apparatus, divided into five portions, and the ^{35}S-radioactivity was determined in each lobe. The right lower and left upper lobes of the lung contained the highest radioactivity while the right middle lobe contained the lowest. Although the distribution of radioactivity was not uniform, intratracheal instillation did provide a reproducible method of exposure.

The second experimental technique was an *in vivo* rat lung study. The technique was a modification of the method of Enna and Schanker;[9] 100 μl of an isotonic sucrose solution containing the radiolabeled compound to be investigated was injected into the lungs via a tracheal cannula. One minute before the end of the absorption period, removal of the lungs was begun. At the end of the absorption period, the blood supply to the lungs was quickly severed. The heart-lung block and a portion of the trachea with the tracheal cannula attached were removed. The lungs and trachea were immediately prepared for the determination of radioactivity remaining in the lungs by the oxygen combustion method.

Sulfate Particulates

Sulfur dioxide, sulfur trioxide, sulfuric acid and sulfate salts comprise a family of compounds that exist as gaseous and particulate pollutants in the atmosphere. Oxidation of pure sulfur dioxide (SO_2) is slow, but the rate of oxidation is increased by light, oxidants such as nitrogen dioxide and ozone, and metallic oxides, which act as catalysts for the reaction. The acids formed react with particulate matter or ammonia to form salts. Most of the sulfur in the atmosphere comes from stationary sources including electric power production. Since the mid 1970s, the automobile catalytic converter has become a new source of atmospheric sulfur compounds. The catalytic converter is capable of oxidizing up to 80% of the sulfur in the fuels to sulfuric acid. Due to its chemical nature, the residence time of sulfuric acid in the atmosphere is probably quite short. Recent studies by atmospheric chemists now point to ammonium sulfate as the predominant sulfur species.

Fly ash produced by coal-fired power generating plants is a ubiquitous source of aerosols containing trace elements. Some elements such as cadmium, lead and nickel are concentrated 100- to 1,000-fold in respirable particles compared to the average concentration in fly ash.[11] These particles may serve as nuclei for sulfate particulates or may be inhaled simultaneously with sulfate aerosols.

Sulfate aerosols appear to be better correlated with human health effects than the parent pollutant, SO_2. The sulfate particulates are complex mixtures of sulfate salts. Exposure of guinea pigs to sulfate aerosols produces a bronchial constriction similar to that of histamine.[12-14] Ammonium and zinc ammonium sulfate are more potent than SO_2, suggesting that sulfate salts may be the most hazardous portion of the sulfur burden of the atmosphere.[15] Previously, we found that sulfate and nitrate salts released all of the histamine stores of guinea pig and rat lung fragments.[16] Thus, the bronchial constriction may be due to histamine release. Histamine release from lung mast cells appears to be dependent on the flux of sulfate ions. Therefore, we wished to measure the removal of sulfate ions in the presence of a number of cations to see if the differing biopotency of sulfate salts could be correlated with their removal flux and histamine release.

Typical experiments depicting the removal of $[^{35}S]$-sulfate ions from the airway of the perfused rat lung are shown in Figure 1 for doses of 0.01 and 0.05 μmol Na_2SO_4. From such curves the initial first-order rate

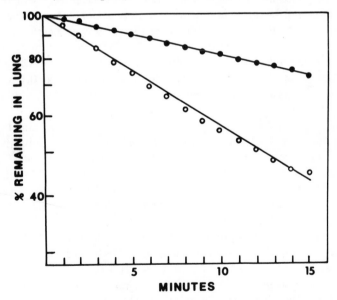

Figure 1. Removal of $[^{35}S]$ sulfate ions from the isolated, ventilated and perfused rat lung. Solid circles, 0.01 μmol of sodium sulfate; open circles, 0.05 μmol.[17]

constants and half-life ($t_{1/2}$) were calculated. The rate of removal of 0.01 μmol Na_2SO_4 was slower than that of larger doses. At doses of 0.05 μmol Na_2SO_4 or greater, the rate was constant. Ammonium ions accelerated the removal of sulfate ions at doses of 0.01 μmol but not at greater doses. The $t_{1/2}$ of 0.01 μmol sodium sulfate was 34.6±5.4 minutes, compared to 16.7±2.8 minutes for 0.01 μmol ammonium sulfate.[17]

To investigate the absorption of sulfate ions in the presence of metallic cations, injections of 100 μl isotonic sucrose solution were used which contained 0.1 μmol/lung ^{35}S-sodium sulfate and 0.1 μmol/lung of the heavy metal cations as their chloride salt. Absorption was allowed to proceed for 30 minutes. Aliquots were collected from the lung effluent each minute and the ^{35}S-radioactivity determined. Salts studied in this manner were cadmium, cobaltous, ferric, manganous, mercuric, nickelous and zinc chloride.

All the heavy metal cations, except manganous ion, enhanced the intratracheal absorption of sulfate ions.[18] Sulfate ion removal followed first-order kinetics independent of the cationic species present. The initial first-order rate constants were calculated from the rate of removal of sulfate ions over 30 minutes or approximately four half lives.

The effects of cations on sulfate ion absorption *in vivo* were similarly studied. The trace metal cations were present in concentrations of 0.1, 1.0 or 10.0 nmol as their chloride salt. As was demonstrated in the perfused lung, the removal of sulfate ions *in vivo* was accelerated in the presence of certain cations (Figures 2 and 3). The dose of cations at which maximum augmentation of sulfate absorption occurred differed from cation to cation. Co^{2+} ions produced maximal effect at 0.1 nmol/ lung. This dose would be equivalent to the deposition of 5.9 ng of Co^{2+} ions on inhalation.[19]

Cd^{2+}, Ni^{2+} and Hg^{3+} cations were of similar potency (1.0 nmol/lung or doses of 112.4, 58.7 and 200.6 ng/lung of these cations, respectively). Fe^{3+} and Zn^{2+} cations were the least effective reaching maximal effect at 10.0 nmol/lung, 558.0 and 654.0 ng/lung, respectively. Mn^{2+} cation, 1 nmol/lung, failed to alter sulfate ion absorption. These levels of cations tested fall within the ranges for the maximum inhaled amounts of metals by man, as reported by Schroeder.[20]

To investigate the absorption of sulfate ions under varying pH conditions, the isotonic sucrose solution containing ^{35}S-Na_2SO_4 was adjusted to a known pH, from 4.4 to 9.4 with either HCl or NaOH, prior to injection. The absorption of sulfate from the rat lung *in vivo* was enhanced at pH values, departing from physiological values (Figure 4). Under basic conditions, the maximum enhancement over control values was calculated to be 26.9±4.4% at pH 9.40. In acid solution, a maximum enhancement

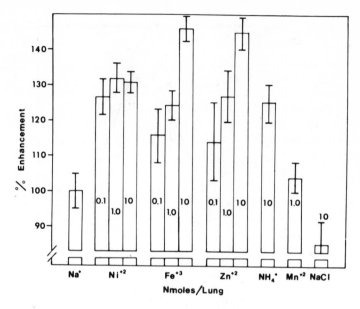

Figure 2. Effects of metallic cations at nanomole doses on the pulmonary absorption of ^{35}S-sulfate ions in the rat. All cations were studied in the presence of 1.0 nmol ^{35}S-sodium sulfate.

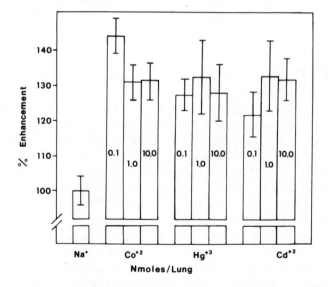

Figure 3. Effects of metallic cations at nanomole doses on the pulmonary absorption of ^{35}S-sulfate ions in the rat. All cations were studied in the presence of 1.0 nmol ^{35}S-sodium sulfate.

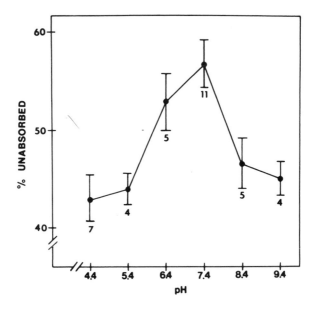

Figure 4. Percent sulfate ion unabsorbed by the rat lung after 30 minutes under varying pH conditions (1.0 nmol ^{35}S-sodium sulfate, specific activity of 0.014 μCi/ mmol). Each point represents the mean ± s.e. The number of animals used at each pH value is given.[19]

of 32.4±5.6% was observed at pH 4.40.[19] Amdur[21] has shown that sulfur dioxide gas produced a lesser irritant response, measured as increased airway resistance in guinea pigs, than the same sulfur equivalent present as sulfuric acid mist, 0.7 μm Mass Median Diameter (MMD).

To investigate the possible effects of a $NiCl_2$ aerosol exposure on sulfate ion removal, rats were exposed to three different concentrations of $NiCl_2$ aerosol with an MMD of < 2μ. Exposure of rats to a $NiCl_2$ aerosol of 480 μg Ni/m^3 for two hours prior to the determination of sulfate ion absorption led to a 12.0±2.7% enhancement of absorption. At the end of the exposure, 856±60 ng Ni^{2+} were present in the lungs of the test rats. Exposure to aerosol concentrations of 113 and 279 μg Ni/m^3 had no significant effect on the absorption.[19]

In summary, sulfate removal from the lung *in vivo* or *in vitro* has been shown to be diffusionally controlled above a dose of 0.05 μmol/ lung. The ability of hydrogen ion concentration and heavy metals, given as intratracheal instillation *in vitro, in vivo* or as an aerosol to test animals, was demonstrated to enhance the absorption of sulfate ions. Mn^{2+} failed to affect the absorption process. In all systems tested there was a positive correlation between the irritant potential of a specific sulfate salt, as reported by Amdur, and the rate at which sulfate ions were cleared from the lung.

Although most heavy metals are known to exert biological effects through combination with sulfhydryl groups, they also combine with hydroxyl, carboxyl, imidazole and amine groups.[22,23] These interactions can lead to membrane changes. The transport of ions across absorbing or excreting membranes has been shown to be sensitive to the action of heavy metal ions.[24,25]

Aerosolized Pesticides

Another class of atmospheric particulates that may have a profound influence on the normal physiological processes of the lung is aerosolized pesticides. Paraquat, a widely used herbicide, has a specific cytotoxicity to mammalian lungs.[26] Acute intoxication is characterized by the development of pulmonary fibrosis.[27,28] After administration of paraquat to animals, the lung has a high initial concentration and retains paraquat.[29,30] This retention appears to be related to the development of lung damage. The mechanism of retention of paraquat by the lung is, at present, not understood. An energy-dependent accumulation of paraquat in lung slices of a number of species has been demonstrated and may account for the retention of paraquat in the lungs of many species.[31] Rat and human lung slices have similar properties for paraquat uptake, making the use of the rat lung an acceptable model for man. Diquat, a herbicide closely related in structure and properties to paraquat, is not actively accumulated by lung slices.[32]

The uptake and distribution of paraquat and diquat in the isolated, ventilated and perfused rat lung (IVPL) were studied by the application of the herbicides to either the capillary or airway surfaces.[33] In Figure 5, a typical experiment is shown representing the uptake of ^{14}C-diquat from the airways mimicking an inhalation exposure. Initially, the ^{14}C was removed rapidly but slowed with increasing time. A semilogarithmic plot of the percent of the radioactivity remaining in lung with time revealed a biexponential decay, which suggested at least two phases of removal of diquat from the airways. The average decay constants for the fast and slow components were calculated from six experiments to have a $t_{1/2}$ = 3.86 and 75.03 minutes, respectively.

A typical paraquat experiment is shown in Figure 6 following intratracheal administration of ^{14}C-paraquat. There was a rapid accumulation of ^{14}C in the lung, and then a slow steady release of ^{14}C into the perfusate was observed. The paraquat efflux rate data could be resolved into two component functions, thus implying that paraquat was absorbed

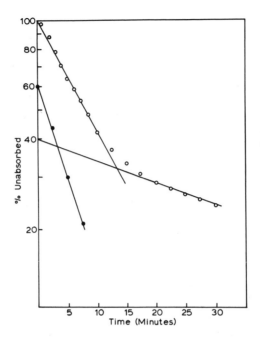

Figure 5. Absorption of diquat from the airways of the isolated and perfused rat lung following an injection of 1.86 nmol/lung ^{14}C-diquat in 100 μl isotonic sucrose (pH 7.4). The rat lung was perfused with a modified Tyrode's solution containing polyvinylpyrrolidone (average molecular weight 40,000; 35 g/l) to maintain a physiological oncotic pressure. The experimental values (○) and the calculated initial disappearance curve (●) are shown.[33]

by two distinct processes by the lung. The average decay constants for the fast and slow components had $t_{1/2}$ = 2.65 and 355.98 minutes, respectively. The slow removal of paraquat from the lung might be attributed to the energy-dependent process that has been proposed to explain the accumulation of paraquat in slices of rat lung. Diquat, on the other hand, is not actively accumulated by lung slices, is not retained by the lung *in vivo* and does not damage the lung. The rapid efflux of accumulated ^{14}C-diquat indicates that the lungs would not be a major site for storage of this compound for any significant period *in vivo*. However, the lungs do play a role in storage of paraquat.

Two pools of paraquat storage were detected from the rates of efflux of ^{14}C from the lung following intratracheal instillation. The second kinetic process was much slower with paraquat than with diquat. On the other hand, perfusion of the pulmonary vasculature with paraquat or diquat resulted in similar uptake and release rates. A major fraction was not retained on perfusion with either compound. The energy-dependent

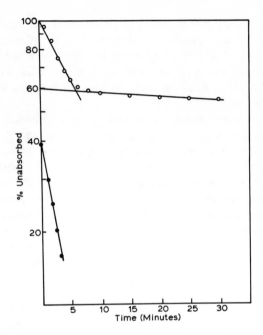

Figure 6. Absorption of paraquat from the airways of the isolated and perfused rat lung following an injection of 1.86 nmol/lung in 100 μl isotonic sucrose (pH 7.4). The rat lung was perfused with a modified Tyrode's solution containing polyvinylpyrrolidone (average molecular weight 40,000; 35 g/l) to maintain a physiological oncotic pressure. The experimental values (O) and the calculated initial disappearance curve (●) are shown.[33]

uptake of paraquat is most likely associated with the airway rather than the capillary side of the lung.

DISCUSSION AND CONCLUSIONS

The effect of heavy metal cations in enhancing the absorption of sulfate ions from the airways of the IVPL into the lung vasculature was studied. The cations investigated included Cd^{2+}, Co^{2+}, Hg^{3+}, Ni^{2+}, Fe^{3+}, Mn^{2+} and Zn^{2+}. The cations increased the sulfate ion absorption from the rat IVPL 199-264% as compared to absorption in the presence of Na^+ ions. An exception was Mn^{2+}, in which absorption did not differ significantly from the Na^+ control. Sulfate ions introduced into the vasculature had the same volume of distribution and mean transit time within the lung as blue dextran, a compound unlikely to leave the intracellular space. Therefore, sulfate ion absorption in the rat IVPL appeared to be unidirectional.

The ability of heavy metals, given as intratracheal instillation *in vivo* or as an aerosol to test animals, was demonstrated to enhance the absorption of sulfate ions. Mn^{2+} failed to affect the absorption process. In all systems tested there was a positive correlation between the irritant potential of a specific sulfate salt[14] and the rate at which sulfate ions were cleared from the lung.

Although most heavy metals are known to exert biological effects through combination with sulfhydryl groups, they also combine with hydroxyl, carboxyl, imidazole and amine groups. These interactions can lead to membrane changes. The transport of ions across absorbing or excreting membranes has been shown to be sensitive to the action of heavy metal ions. The enhanced sulfate ion absorption observed in the experiments presented here is probably related to membrane changes resulting from these interactions.

Acidic aerosols of sulfuric acid are produced by several environmentally important processes including stationary power generating plants and catalytic converters on motor vehicles. Sulfate ion absorption was enhanced under both basic and acidic conditions. Since both acidic and basic ionizable groups exist on all membranes, a diffusionally controlled process, such as sulfate ion absorption, could well be influenced by the pH of the local extracellular environment. Little is known about the effects of the inhalation of acid aerosols from the environment on the pH of the localized film of water in the airway. We have no knowledge of the buffer capacity of the airway lining fluids, and, thus, large changes in pH may result from the inhalation of small amounts of acid. The pH of a sulfate aerosol will therefore have an effect on sulfate ion removal.

Data presented here suggest a correlation between the rate of sulfate ion absorption from the mammalian lung and the reported bronchoconstricting action of some sulfate salts. Clearly, the rate of sulfate ion absorption is influenced by the cationic species present and the pH of the surrounding extracellular environment.

Herbicides and other pesticides may enter the human body by inhalation, dermal absorption and ingestion. Diquat and paraquat represent two herbicides of very similar physical, chemical and biochemical properties but of vastly different toxic action to the lung. The energy-dependent uptake of paraquat by lungs of several species including man may represent an important process for the accumulation of other potentially toxic compounds by the lung. Inhalation may represent a distinctly different hazard than ingestion or dermal absorption.

Two pools of paraquat storage were detected from the rates of efflux of ^{14}C from the lung following intratracheal instillation. The second kinetic process was much slower with paraquat than with diquat. On the other hand, perfusion of the pulmonary vasculature with paraquat or diquat resulted in similar uptake and release rates. A major fraction was not retained on perfusion with either compound. The energy-dependent uptake of paraquat is most likely associated with the airway rather than the capillary side of the lung. Type II pneumocytes have been suggested as the major site of paraquat storage and paraquat is thought to produce pulmonary damage by acting on these cells. The long-term retention of intratracheally instilled paraquat found in our experiments supports this hypothesis.

In conclusion, while kinetic measurements in themselves provide no direct mechanistic proofs, the use of kinetic parameters to characterize the removal process and the effect of chemical structure on the kinetic parameters has allowed the development of testable models. In predicting the effects of the inhalation of toxicants in man, of particular importance is the determination of the mode of removal, *e.g.*, diffusional, facilitated or active transport. Studies directed toward such determinations are possible from the experimental approaches discussed.

Lastly, the removal of materials, especially electrolytes, from the airways may represent important normal physiological mechanisms for the maintenance of the integrity of the lung. Pathophysiological changes may accompany disease processes affecting the lung and may be detectable from a greater knowledge of the biochemistry and biophysics of toxicant removal.

REFERENCES

1. Enna, S. J., and L. S. Schanker. "Absorption of Saccharides and Urea from the Rat Lung," *Am. J. Physiol.* 222:409 (1972).
2. Enna, S. J., and L. S. Schanker. "Absorption of Drugs from the Rat Lung," *Am. J. Physiol.* 223: 1227 (1972).
3. Enna, S. J., and L. S. Schanker. "Phenol Red Absorption from the Rat Lung: Evidence of Carrier Transport," *Life Sci.* 12:231 (1973).
4. Taylor, A. E., and K. A. Garr. "Estimation of Equivalent Pore Radii of Pulmonary Capillary and Alveolar Membrane," *J. Physiol.* 218:1133 (1970).
5. Burton, J. A., and L. S. Schanker. "Absorption of Antibiotics from the Rat Lung," *Proc. Soc. Exp. Biol. Med.* 145:752 (1974).
6. Burton, J. A., and L. S. Schanker. "Absorption of Corticosteroids from the Rat Lung," *Steroids* 23:617 (1974).
7. Normand, I. C. S., R. E. Oliver, E. O. R. Reynolds and L. B. Strang. "Permeability of Lung Capillaries and Alveoli to Non-electrolytes in the Foetal Lamb," *J. Physiol. London* 219:303 (1971).
8. Taylor, A. E., A. C. Guyton and U. S. Bishop. "Permeability of the Alveolar Membrane to Solutes," *Circ. Res.* 16:353 (1965).

9. Burton, J. A., and L. S. Schanker. "Pulmonary Absorption of Aerosolized Solutes," *Fed. Proc.* 30:447 (1971).

10. Niemeier, R. W., and E. Bingham. "An Isolated Perfused Lung Preparation for Metabolic Studies," *Life Sci.* 11:807 (1972).

11. Davison, R. L., D. F. S. Natusch and J. R. Wallace. "Trace Elements in Fly Ash: Dependence of Concentration on Particle Size," *Environ. Sci. Technol.* 8:1107 (1974).

12. Nadel, J. A. In: *Proc. Second Int. Symp. Inhaled Particles and Vapors*, C. N. Davies, Ed. (Oxford, England: Pergamon Press, 1966), p. 5.

13. Amdur, M. O., R. Z. Schultz and P. Drinker. "Toxicity of Sulfuric Acid Mist to Guinea Pigs," *AMA Arch. Ind. Hyg. Occup. Med.* 5: 318 (1952).

14. Amdur, M. O., and D. Underhill. "The Effect of Various Aerosols on the Response of Guinea Pigs to Sulfur Dioxide," *Arch. Environ. Health* 16:460 (1968).

15. Amdur, M. O., and M. Corn. "The Irritant Potency of Zinc Ammonium Sulfate of Different Particle Sizes," *Am. Ind. Hyg. Assoc. J.* 24:326 (1963).

16. Charles, J. M., and D. B. Menzel. "Ammonium and Sulfate Mediated Release of Histamine," *Arch. Environ. Health* 30:314 (1975).

17. Charles, J. M., W. G. Anderson and D. B. Menzel. "Sulfate Absorption from the Airways of the Isolated Perfused Rat Lung," *Toxicol. Appl. Pharm.* 41:91 (1977).

18. Charles, J. M., and D. B. Menzel. "Heavy Metal Enhancement of Airway Sulfate Absorption in the Perfused Rat Lung," *Res. Comm. Chem. Pathol. Pharmocol.* 15:627 (1976).

19. Charles, J. M., D. E. Gardner, D. L. Coffin and D. B. Menzel. "Augmentation of Sulfate Ion Absorption from the Rat Lung by Heavy Metals," *Toxicol. Appl. Pharm.* In Press.

20. Schroeder, H. A. "A Sensible Look at Air Pollution by Metals," *Arch. Environ. Health* 21:798 (1970).

21. Amdur, M. O. "The Impact of Air Pollutants on Physiologic Responses of the Respiratory Tract," *Proc. Am. Phil. Soc.* 14:3 (1970).

22. Passow, H., A. Rothstein and T. W. Clarkson. "The General Pharmacology of the Heavy Metals," *Pharmacol. Rev.* 13:185 (1961).

23. Ramachandran, L. K., and B. Witkop. "The Interaction of Mercuric Acetate with Indoles, Tryptophan and Proteins," *Biochemistry* 3: 1603 (1964).

24. Bank, N., B. F. Mutz and H. S. Aynedjian. "The Role of Leakage of Tubular Fluid in Anuria Due to Mercury Poisoning," *J. Clin. Invest.* 46:695 (1967).

25. Knauf, P. A., and A. Rothstein. "Chemical Modification of Membranes. I. Effects of Sulfhydryl and Amino Reactive Reagents on Anion and Cation Permeability of the Human Red Blood Cell," *J. Gen. Physiol.* 58:190 (1971).

26. Clark, D. B., T. F. McElligot and E. W. Hurst. "The Toxicity of Paraquat," *Brit. J. Ind. Med.* 23:126 (1966).

27. Kimbrough, R. D. and T. B. Gaines. "The Toxicity of Paraquat to Rats and its Effects on Rat Lung," *Toxicol. Appl. Pharmacol.* 17: 679 (1970).

28. Brooks, R. E. "Ultrastructure of Lung Lesions Produced by Ingested Chemicals. I. Effect of the Herbicide Paraquat on Mouse Lung," *Lab. Invest.* 25:536 (1971).

29. Sharp, C. W., A. Ottolenghi and H. S. Posner. "Correlation of Paraquat Toxicity with Tissue Concentrations and Weight Loss of the Rat," *Toxicol. Appl. Pharmacol.* 22:241 (1972).

30. Litchfield, M. H., J. W. Daniel and S. Langshaw. "The Tissue Distribution of the Bipyridylium Herbicides Diquat and Paraquat in Rats and Mice," *Toxicology* 1:155 (1973).

31. Rose, M. S., L. L. Smith and I. Wyatt. "Evidence for Energy-Accumulation of Paraquat into Rat Lung," *Nature* 252:314 (1974).

32. Clark, D. G., and E. W. Hurst. "The Toxicity of Diquat," *Brit. J. Ind. Med.* 27:51 (1970).

33. Charles, J. M., M. B. Abou-Donia and D. B. Menzel. "Absorption of Paraquat and Diquat from the Airways of the Perfused Rat Lung," *Toxicology* 9:59 (1978).